THE APPLE CORPS GUIDE TO THE WELL-BUILT HOUSE

The Apple Corps Guide to the Well-Built House

Jim Locke

Foreword by Tracy Kidder

A Richard Todd Book
HOUGHTON MIFFLIN COMPANY · BOSTON
1988

For Rachel, Sara, Corinne, and Josh,
and, of course, Sandy

Library of Congress Cataloging-in-Publication Data

Locke, Jim.
The Apple Corps guide to the well-built house.
"A Richard Todd Book."
1. House construction — Amateurs' manuals.
2. Dwellings — Remodeling — Amateurs' manuals. I. Title.
TH4815.L63 1988 690'.837 88-9370
ISBN 0-395-43042-9
ISBN 0-395-47808-1 (pbk.)

Printed in the United States of America

Drawings by Raymond L. Porfilio, Jr.

Book design by Robert Overholtzer

S 10 9 8 7 6 5 4 3 2 1

Contents

Part Three
Finish Work

Acknowledgments

I am grateful so many people transcended my Little Red Hen attitude and helped this book get from the field to the table. My grandfathers, each in his way, started me in my honest trade and made me think I could write, too. Sandy Warren urged me past my novice's frustrations with her clear mind and good spirits, and her love. Our children, Rachel, Sara, Corinne, and Josh, watched this new endeavor with the humor and fond skepticism of true teenagers. Tracy Kidder showed me writing could open my eyes and mind, and I thank him for his touching foreword. My parents and siblings offered encouragement and high standards — a potently productive combination. Dave Warren granted me my first interview, and his interest moved me along.

Alex, Rich, Ned, Bubba, Donny, Dennis, and Fred tolerated my absences from various job sites, and gave willingly of their trade smarts. Bill Bisbee of Bisbee Bros. Lumber patiently filled in my knowledge of materials. Chris Packard and Harvey Schlesinger got me started with word processing, and Tony Harris rescued me, several times, near the end. Samantha Harris generously loaned me her computer when mine gave out. Linda and Gene Epstein believed this project possible when I didn't.

I got lots of technical information from experts in the sub

trades. Gene Poissant and Norm Nye, for sixteen years our electrician and plumber, have taught me volumes. Ken Denno knows heating systems, and cares about a top-notch installation. Dave Pogue, the zookeeper, produces great drywall jobs and a good time, too. Doug Magee had the courage to speak for all building inspectors. Lynn Rice's thoughtful words on house architecture show her commitment to her profession. My old buddy Bert Willey grudgingly taught me everything he knows about painting, and we played pool for the second half hour.

Jeff Cruikshank helped me get this book off and running. Dick Todd organized my early efforts into some order and, throughout, saw me past the hurdles of my temperament. Ray Porfilio, architect by day, got the drawings just right: informal but informative. Larry Cooper, a genius with a red pencil, made the manuscript cohesive, readable, and orderly. He taught me things that had always eluded me, which took some doing.

Patricia Van der Leun, my agent, was the moving force behind my writing. She push-started a reluctant builder, and the result is this book.

To you all, I'm much obliged.

Foreword

About ten years ago, I set out to fix up an old house by myself. Not knowing most of the names of the parts of a house or of the tools for putting one together, I went like a waif to lumber yards, where I begged for morsels of information. I didn't know I needed a lot more help than that, until I was on the way to the nearest emergency room. I had inserted the blade of a power saw about halfway through the knuckle of my right thumb. I felt like a fool, of course, and as I sat whimpering over the bloody mess, I couldn't get out of my mind the memory of an anthropology text that had stressed the evolutionary importance of the opposable thumb. Looking back, I wish there'd been a book around like this one, to make me understand the complicated thing that I was dealing with, and to make me see the wisdom of setting aside my tools until I had the money to hire professionals, which is what I did, one stiff thumb too late.

One is never entirely sorry to see carpenters leave a job for which one is paying. The first bunch who came to my house did what I thought was pretty good work; that is, it was a lot better than mine. But they wouldn't talk to me much, and after I overheard them making some disparaging remarks about my house, I really couldn't wait for them to go. The next year I hired the improbably named Apple Corps. They actually let me work with them from time to time, not in anything that took skill but in holding the other end of boards, so to speak. Jim Locke writes that homeowners who insist on working with their carpenters make the building more expensive than it has to be, but he didn't tell me then that I was paying tuition. I felt sad when he and his partners departed. It wasn't just that I missed their company. They interested me, the way, I suppose, accomplished craftsmen in any field interest the dabblers. I pursued them for the better part of a year before they agreed to let me watch them build a house, and write about what I saw.

I remember one of Jim Locke's partners saying, one day on that job, that they should put a bell around my neck, so they'd know when I was nearby and could avoid saying more than they wanted me to hear. I think that at one time or another all of the people portrayed in *House* felt uncomfortable about my intrusions into their lives. But none ever tried to get control over what I wrote, and none, I think, tried to hide much. Journalists are lucky when the people they observe turn out to be candid, and even luckier when they turn out to be perceptive besides. Jim Locke was both, almost unfailingly, for me. From the first time I confronted him with my notebook open, he struck me as one of those people who is interested in being honest for the sake of being honest, and also in being accurate. Jim seemed unusually disposed to think about what he did and why he did it, not failing in the meantime to examine the performances and motives of most of the people around him.

In many old pictures, house carpenters wear bow ties and long-sleeved shirts with garters at the elbows. I remember Jim on the job, dressed in less formal work clothes that still managed to convey a similar air of composure. His were always clean and neat. Behind his ear rested a freshly sharpened pencil, which — with one of his molding saws — he had cut down to the size that allowed it to rest most securely there. In my memory, Jim is surveying the house's new foundation, worried, among other things, that it may be more than marginally out of level. If it is, as foundations often are, he'll have to put shims under the first layer of wood. That is common practice and does not make a house unstable. Moreover, once the building is finished, one can't see the shims except from down in the basement, and only then if one looks for them. Jim is not impractically meticulous; he'll accept shims if he has to. But he'd rather not. "Because they look cobby," he says.

Many people who call themselves builders nowadays have never lifted a hammer for pay. Some care about the quality of the buildings they arrange for others to construct, but they can't possibly understand what craftsmanship in building means, the way an actual, scrupulous builder such as Jim Locke does. Long before I met him, Jim had discovered that for him life was much more purposeful if he tried to do things right than if he left what he and his partners call "cob jobs" behind. Jim cares about the quality of the joinery in the parts of buildings that get covered up, as well as in the surfaces. Watching Jim and his partners at work, I gradually realized that they managed their time efficiently and calmly, in a much more adult way than I had ever done. I'd sawed halfway through my thumb partly out of clumsiness and partly because I'd been in too much of a hurry to put a sharp blade in my power saw, and had ended up wrestling with the tool in order to make it cut. Jim and the other members of Apple Corps spent a lot of time keeping the construction site neat, arranging lumber piles, setting up tools and workbenches. They spent time to save time, and also their bodies. Building the staircase, Jim stopped at least once every hour, and often more frequently, to sharpen the tools he was using.

As the house arose, he began writing down the names of the owners and builders, his own included, of course, on shingle shims and on the back of moldings. He nailed these up in obscure corners, where no one except another carpenter would be likely to find them. Evidently, he was writing notes to the future, for carpenters who hadn't been born yet but would someday come to work on the house. Later, when the builders had finished and my book was published, Jim told a reporter that the house they'd built would probably survive a lot longer than the book I'd written. I was secretly annoyed, most of all because I knew that, barring fire or bulldozers, he was probably right. Now Jim has ventured into this fragile medium, of paper rather than wood. He went about making this book in the same way he goes about building a house: he thought it all out ahead of time; he set up a production schedule; and, most unlike a writer, he stuck to his plans. He was determined not to leave cob jobs behind for repair later on, but he got better as he went along, and so, it pleases me to say, he had to go back and fix some of his early work after he got to the end of the job. This book is entirely his, and he has left his mark on it.

I remember bringing Jim a sketch for a garden shed, intended for my yard. He looked at it and said, "This isn't what you want." He made his own sketch, and built from that. I'm still not sure he would have agreed to make the thing I'd had in mind. Jim has opinions about everything to do with building. These have made for some lively discussions when he has come up against equally opinionated customers. But certitude is what one wants in a builder, especially in one dispensing advice. In this book, Jim acknowledges his prejudices, and he describes many of the alternatives that he doesn't like. If a reader ends up doing what Jim wouldn't do himself, the reader at least will know what he is getting into. Jim has laid out, step by step, the process of building a house. He has not written a manual for doing it yourself, though the book would help anyone who planned to make that error. His book conveys a much greater conceptual sense of what is involved in creating a house than any manual I have seen. It lays out very clearly the terminology and psychological dilemmas of construction. Anyone who reads it should come away equipped to talk with builders and to begin to understand them.

I once heard Jim say, "My feeling about trim is, you don't want to have to look at it more than once. It should be securely there, it should define the openings, but it shouldn't be gimmicky." He has applied this self-effacing principle of aesthetics to his prose. His obvious intention isn't flashiness, but gracefulness and explanation. Jim also once remarked, "The interesting parts are the edges, where things come together." In antique parlance, the term "joinery" described the most difficult parts of carpentry. In good joinery, everything fits together snugly. This is a book by a skillful, principled joiner.

TRACY KIDDER

Part One

Planning and Paperwork

1

Build a Good House

You want to build a house, and I want you to build a certain *kind* of house. Not necessarily a colonial or a contemporary, but a solid and enduring building that fits you and looks good. I don't advocate any particular style, though I know some materials and designs last longer than others. I don't urge that you build a mansion or a work of art, though the suggestions in this book apply to any building. I do urge you to pay for a house you and your heirs will likely never use up.

When you undertake to cover up a piece of the planet with something you build, you are duty bound to build it well. A new house can be made to last hundreds of years. Even a poorly constructed one may endure several decades, but it will always cost its owner money and aggravation, and society will inherit the task of supplying it with resources or cleaning it up if it fails utterly. Most houses don't get the best of everything because it costs too much. No matter what you build, though, your new house will cost plenty, and will satisfy you only if you spend your money wisely.

I am a partner in Apple Corps, a small custom-building firm. ("Custom" means for you, as opposed to "spec" building, which is for whoever comes up with the cash.) We are five partners and several employees. We do most of the actual construction of our projects with our own hands. Our work is mainly single-family houses, and is divided between building new ones and remodeling older ones. This book proceeds as if you were building from scratch, but much of it also applies to renovations.

Though I write this book by myself, I am informed by the attitudes and experiences of the whole company. We have customers from many income levels, and build accordingly. We think about our houses, our customers, and the business of building all the time. I know what I've learned will help you.

A house is lots of pieces put together with a lot of work. The sum, though, can be more or less than the parts. You've been to houses, and especially to apartments and condos, that just don't look good or work well. The square footage is there, the appointments are nice enough, sometimes the design and layout seem all right. But something is wrong. Look closely — the drywall joints are visible, the baseboards don't quite meet in the corners, and the kitchen counters barely fit the walls behind them. The security, the emotional safety, a house should instill in you is missing, because you just aren't sure whether the structure is sound. Are the invisible but necessary supports and fasteners all correctly in place? You may not know what should be back there, but you want it to be *right*. You shouldn't have to worry about your house!

A demoralizing fact of life about houses is that they usually look better than they are constructed. Spec, or production, builders spend their money on "features," the visible parts the real estate agents advertise and buyers look for and know something about. The necessary and permanent structure of the house is rarely discussed or understood by buyers — or sellers either, for that matter.

So all the important structural and weatherproofing jobs can be finished with little scrutiny from the public. The framing, roofing, windows, doors, and siding are often all done before the first few hardy buyers appear. The money the builder saves, if it comes from skimping, rarely affects the salability of the house. But shoddy work hidden behind the finished surfaces will spell trouble in a few years. Here's an example of what I mean.

The exterior of a house must protect itself and you from the elements. Every exterior woodwork joint must be closed — that is, the ends of the boards must be out of the weather. The trim that runs around the roof just below the shingles is a case in point. The outside walls of new houses are usually built lying flat on the rough floors and tipped up into place. Production builders will add the roof trim to the wall while it is still on the floor. Trouble is, it's too hard to get a tight-fitting joint at the lower end of a trim board installed that way. So you get a squirt of caulk and a paint job instead. This won't last but will look fine from the ground long enough for you to buy the place.

Manufacturers constantly come up with products, such as aluminum roof trim and metal chimneys, that diminish the skill necessary to get a job done. Close tolerances and exact fitting are not required. Cheaper, less skilled help can trim the same house for less money. The object is that of the assembly line: break down each job into parts that someone can be easily (and inexpensively) trained to do. If the result is less than perfect — hey, you saved some bucks, right?

Wrong. Your short-term saving may be substantial, yet if you keep your house even a decade, quality materials and workmanship will save the difference in lower maintenance. From then on, your house will outlast and outshine its skimped-on neighbor, and this will be reflected in the resale price. And all along the way, the better house will return to you lower running expenses, fewer problems, and more satisfaction.

Once you decide on a well-built house, you have a few choices. You can buy an older house or a new spec house, if you can find one put together properly. You may feel the urge to build from scratch with your own hands or decide to act as your own general contractor. Or you can hire a custom builder and an architect, the most rewarding approach for most people.

If you buy a house already built, you'll have to check it out thoroughly. You must accept the design and layout, the quality of materials and construction, and any wear and tear. If you like the house but not the neighborhood, you're stuck. If it's a new house built on speculation, be aware that your best interests didn't shape the project.

Building a house yourself, with little training and experience, is a formidable task. Building covers many fields and skills, few quickly mastered. You might, given enough time, do a better job than a professional. More commonly, your house project will take your spare time for a decade. Few do-it-yourselfers conquer the skills early enough in their project to lend full value to the money they do spend; what they save on carpenters they lose on resale or upkeep.

But perhaps you can just be your own general contractor. What that means is you will organize the job, order materials for timely delivery, arrange for and supervise the subcontracting, obtain applicable permits and approval when necessary, and generally advise and supervise the workers at each stage of planning and construction. This sounds like a lot of work and it is. The contractor sets the tone for all the workers to follow. Quality comes hard, and he must demand it of those he hires, not take it for granted. This, especially, is difficult for anyone new to the business.

The only practical way a novice can get through it is by being very open-minded about the completion date. General contractors, experienced in scheduling work crews and having earned top-of-the-list status with many subcontractors, still have trouble meeting proposed deadlines. Whether you wish to wield the control you're used to, or wring the best deal possible from each supplier or subcontractor, you should know you will have more trouble. Inexperienced contractors rarely get the best job from their subs or the best service from their suppliers.

I built my first house from a book, having previously tackled only tree houses. The author championed a particular style and method, and I followed dutifully, changing a few areas according to my limited experience. Since I had never built anything like it before, the fact the house is still standing fifteen years later means, I guess, that the author and I made a successful team.

Through the ten years I owned the place, though, I didn't like it. Some of the problems were in my workmanship and some in his design and construction advice. Others appeared in my evolving taste and knowledge of building. I have become more philosophical about the whole project since I sold it, and the fellow who bought it has done much to improve it. All this is not to stop you from attempting to build or oversee the construction of a house yourself. You will surely learn a lot from trying, and you may even save money if you learn enough.

Your house will be made of many parts. If you decide on building a custom one, you will have to choose those parts. You may hire an architect to help you or depend on a builder. Knowing your goals and understanding the implications of your wishes prepare you for the unending list of decisions you will make. I will give you details, details, details. Knowing the details informs your choices. And really, choosing is your main job throughout the project.

I will urge you to invest most of your time before any hole is dug, any nail driven. I'll drag you over and over your plans and specifications until you're as sure of them as you can be. I'll tell you how to hire builders and architects, how they choose their customers, and what kind of customer helps his job along.

I encourage you to hire a custom builder and, in many cases, an architect. A custom builder or general contractor comes with few strings attached because his work history is broad, not repetitive. If you put a decent price on your time, you'll lose money at general contracting your own house. If you have little experience and plan to build your house yourself, look at it as a hobby or a fling. If you seriously want a good house built well, without medical side effects, hire it done. And to do that, you'll have to decipher everything in advance, starting with yourself.

2

Who Are You and What Do You Want?

IT'S IMPORTANT TO decide what you're trying to get by building a new house. I'm talking about goals, for you. Your goals for your house should follow those for your life, and support and sustain them.

If you're retiring, for example, you probably figure on staying in the house for the rest of your life. If you plan to travel a lot, your house should be easy to close up and leave without worry. Perhaps you'll need a wing you can open when the family comes to visit. You'll want a house that's easy to pay for and to get around in. You may want some of the formality of strictly separate rooms.

A midlife couple with a big family and two careers in full song needs another kind of house. Four teenagers and their friends take up a lot of room. The kitchen and eating areas will see near-constant use. A separate, quiet study or office for each of you may be a must. You'll need lots of parking spaces, closets, and telephones.

The way to get your kind of house is to list your goals, starting with the most general, for presentation to your architect and builder. Your list should be broad, not specific. Each time you want to write "den," write "quiet reading area" or "separate hobby space." You want help with the design, so don't make design decisions ahead of time. Your being clear about what you want to end up with is the most help you can give those who design and build for you.

Please don't wait until the meetings, or, worse, the drawings, are under way to figure out what you want. If you do, you will waste time and money on corrections. And quite likely you'll compromise your goals so as not to interrupt the flow of work or ruffle the feathers of the designer. So figure out what you're after ahead of time, and be willing to stick to your mission.

If you are a couple, try for everyone's sake to get a few things straight before you call the builder. Having a house built severely tests the bonds of marriage, and many don't survive the strain. Every builder knows of a couple divorcing after he finished their new house. I'm not a psychologist, but I have a few notions of why this happens.

Certainly, one problem can come from trying to solve marital problems with the house. Saving the marriage by building a house is about as likely as saving it by having a baby. It seems reasonable that more space for the spouses or kids, a specialized room or two, or a bigger yard would alleviate some of the tensions of family life. Obviously, if tight quarters is the only problem, more space is the cure. But that's seldom the whole story. In the end, the family must convene sometimes, and the dinner table can witness scenes all the sewing rooms in the world won't smooth over.

Underlying conflicts between partners over their contrasting aspirations for the house keep coming out all through the job. No builder or architect can help you when you can't resolve your own differences. In meetings, I'm often called on to take sides. I sometimes have a preference, but choosing means agreeing with one party at the expense of the other. I usually try to mediate these disputes by alternately favoring each side, but it's tiring, and not always honest. If this venue is the only outlet for your differences, though, go ahead and use it. Resolving the issue is most important.

Another kind of thinking is in order now, too. Figure out how

nt react to putting your immediate destiny in the hands
nger, the builder. One thing that might happen is some
l male rivalry. Imagine an early meeting with your
getting acquainted and talking about houses. The husband wants to prove, to the wife or to the builder, that he can take charge of the situation. Dodging and feinting, he leads the conversation in circles, trying to dizzy his imagined opponent. If a subject he knows little about comes up, he waits for the wife to ask a question. Then, as soon as he hears the answer from the builder, he acts as if he knew it all along. If this is you, and you can't control yourself, try to find a patient builder who's bigger than you, and good luck.

The main reason for all this trouble is the huge number and potent nature of choices the new house will demand of you. If yours is a normal project of four or five months of planning and as many of building, you'll be facing several decisions per day for almost a year. Some of these choices are bigger than others, but all cost money and showcase your personality. Allocating money for things is a common battleground in any relationship, and here you'll be dispensing large sums day after day. And the deep feelings we all have about our houses will be on display, bringing nerves near the surface, unprotected.

Sweet's Catalog is a manual of manufacturers' brochures covering many of the products used in construction. Many architects use Sweet's to choose or recommend options for their clients. Sweet's runs many thousands of pages, in sixteen dictionary-size volumes. I mention it to show how many things you'll have to decide on, never mind how they are put together.

You must first make several big decisions, like where to build and how much you can spend. You will resolve perhaps a dozen only slightly less important issues, like who will draw and build, how to finance, general style and layout. And then probably two thousand other decisions, large and small. The baseboards inside the closets may be different from those in the rooms — is that OK? There are hundreds of cabinet knobs from twenty manufacturers, from different stores; all styles don't come in all finishes; some may be available only by mail. Will they show up on time? Should we get . . . choose one now! And that's only cabinet knobs.

No purchase displays your image and aims in life as unfailingly as your house. You will be gripped by the magnificent importance of every selection you make, especially for your new custom-built house, about which you make most of the decisions. Here lies one advantage of buying an already-built house: most of the choices were made before you got there. It's easier to accept a poorly placed window or inadequate kitchen counters if you didn't order them yourself.

Each hour you spend figuring out what you want and how to do it before you start will be worth two, or maybe ten, to you once the project is under way. It is far easier for everyone to agree to guidelines and constraints ahead of time than when the crew is waiting around for instructions. Even if you can prevail in most of these too late contretemps, the finished house will exhibit the argument, not the winner.

Like raising children, but compressed in time, hiring a house built exposes your habits and foibles to strangers and to yourself. The scrutiny is rewarding, sure, and stressful. You will be forced by the task into a wider self-knowledge. You'll be intimate with your new house, and the planes and angles will be filled with stories. It will be your house.

Before you start planning, set out to understand some of the influences on your decisions. I don't mean the underlying ones like culture, current economic conditions, or your family structure, but the more direct factors: advice from builders, architects, and friends; the place you live in now; and money.

You will get a lot of advice before you start hiring designers and builders. Most of it is harmless enough. If what you hear appeals to you, ask the architect or builder about it in early discussions. Advice becomes useful when it makes you wonder why you choose as you do. It is onerous when you follow it to win approval.

Your impressions of friends' houses and those you see in magazines and from the street are most useful if you can tell why you react a certain way. Do you hearten to the reassuring formality of classical symmetry? Are you liberated by the whimsy of a tower? Did you fall for a porch you saw in a clothing ad? Why? Completed projects are often easier to respond to than one of your own blueprints. Try to connect your reaction to an emotion you can talk about. That's a big help to your designer and builder.

Building part of your house to shelter a dependent relative is a poor plan until you check with that relative. Simulating your grandparent's house or room allows you a heady opportunity to re-create prized childhood memories. Memories are tricky things to build, though, so knowing the qualities of your vision will be the most helpful information for your designer. It's important for you to be open-minded about proposed designs — you may find a totally new layout you like better.

Another large and often-overlooked influence on your new design is your present quarters. After enduring a tiny hallway kitchen for years, you can't wait to move into your new 20-by-36 country kitchen. Later, you find you get worn out shuttling around it. Similarly, a third bathroom might be redundant in your present house but necessary to the layout of your new one. A compromise often yields a better long-term solution. The commanding influence of your current life is sometimes hard to acknowledge. Here your architect's objectivity proves his worth.

Money, though, is the setter of standards for your house, the sieve through which all your decisions will be forced. In most cases, you will want things you can't afford. The more expensive windows are not always the better ones, but often enough they are. Each item in your house is going to be available in a range of prices, which can vary according to the style and quality. Today's decisions can affect the ones you'll make in a month. Some finish details will be forced on you by choices you make early in the planning. This thicket is daunting, and your builder or architect must help you through. Every thing in a house costs money.

Ask people who have owned their new house a few years and almost all will tell you they're glad they built when they did. Inflation and appreciation have ensured their financial satisfaction. Those who spent a little more for a better job are the best rewarded, as their house constantly returns their investment. Hiring the lowest bidder guarantees you will get a builder, suppliers, and subcontractors who are cost-conscious. This means *their* cost, of course, not just yours. Every decision about your house will be made with an eye toward the cheapest material or method.

There is no economy in paying $150,000 for a house that doesn't fit your family when $15,000 more would make it really work. Or the extra 10 percent could buy you just the same house, but built better by a more conscientious builder. Again, the money is well spent. Fixing something always costs more than building it right the first time, particularly big-ticket items like leaky basements, poor windows, or cheap siding shedding its finish. It's always the last few percent of the budget that causes the most anxiety, since the end of the spending spree is in sight. But it works against you to be stingier with that last few than with the first few — they all contribute to your satisfaction.

Your project's budget should be 3 to 6 percent less than the total amount you can spend on the house. If you are quite sure you have complete plans that are to your liking, and if you have strong faith in your builder and architect, reserve 3 percent. If any area of the plans, or your confidence in the principals, is in question, save 6 percent of your money for midproject spending on changes, additions, contingencies, and catastrophes. I'm talking about thousands of dollars, and there are several reasons you'll be glad you have it.

Over and over again, customers find some item or feature they want after their contract is signed and their house started. This often is something they talked themselves out of in the early trimming of the budget. Or the reality of the house as it emerges in studs and plywood isn't quite what they thought, so they must change or add a few things. Some want to play their cards close to the vest, and see how they like the progress of the building before they reveal their resources. Whatever the reason, few customers spend as little as they plan.

Architects and builders often recommend the more expensive items when you are choosing. This tendency is somewhat different from that of the car stereo salesman who would have you buy the latest tri-amped active-crossover system. For the salesman, the main advantage of the high price is more profit to him. But a builder must physically and financially stand behind his work. For him, your choosing the higher-quality window means fewer installation and callback problems. He's likely to get better service from a superior manufacturer in case of difficulty. He wants to build a house that he can show off to prospective customers without explaining, "Of course, we wanted to use Ultra-Sash windows but the budget stopped us." And, sure, he's happy to make more money.

The best architect or builder may have trouble helping you visualize what he proposes to build for you. When you finally see the house going up and the finish work going in, you might be surprised or disappointed. There'll be a lot of pressure to keep the job going, and only with money can you get the changes you want. This is rarely an easy time for you, even though your dream house is being built. The job is moving at its own pace at that point, often well outside your control and understanding. So if you have some money reserved, you have some power to effect what you want. If not, all you have is the trouble.

Contingencies and catastrophes are usually the result of items left out of, or poorly defined in, the contract. Everyone's best intentions and experience don't always prevent omissions. Whole categories, like gutters or bath accessories, might be left out, but you still need them. Or an item — linen closet shelving, for instance — is mentioned but not spelled out in the contract. When the time comes to put in the shelving, the contractor is thinking, ". . . with four shelves, two feet apart," and the customer is thinking, ". . . shallow cubbies for washcloths, deep blanket shelves, hangers for brooms . . ." You'll be happy to have money left to solve these problems.

There also exists a gang of expenses often not in the building contract but necessary when you move in. These include telephones, curtains and drapes, mailboxes, kitchen and laundry appliances, and changing your stationery. Do not underestimate how shabby your old furniture and rugs will look in your new house — some of it will have to go. You can spend thousands of dollars outside the contract, but only if you have it. The trick lies in not letting the house itself consume all your available funds.

Your big task now is to establish a budget for your house project. Figure the total amount you can safely spend by talking to your banker, accountant, or financial adviser, not to friends or relatives. Be sure you can afford the payments, but remember, few people lose money on a house they build, unless their local economy takes a permanent nose dive or they are extravagant. You may not want all your assets frozen in real estate, or be able to predict your income for the term of the mortgage. These are personal financial matters you must consider. Draw up a budget amount you will stick to, come what may, and remember the 3-to-6-percent rule.

3

Planning and Hiring

NOW THAT YOU have a basis for thinking about this project, it's time to move along. You have to choose a lot, a builder, an architect or a set of plans, and learn something about getting along with these people and their minions. Your crucial work starts now, for setting up the job right is your best assurance of satisfaction. A sample chronology of the start-up should help you get organized.

Here's a good timetable for a normal construction season. Let's say that in December 1989 you come up with the idea of building. Consider the questions of buying versus building, new versus old, new spec versus new custom, condo versus house, and so on. If you decide on a new custom-built house, load yourself up with facts and advice by talking to everyone you can think of, and everyone they can think of. I mean advice from friends, neighbors, relatives, and the like, not from anyone with something to sell. Consider builders and architects, and collect information on styles, floor plans, and lots.

Next, call or meet some builders. These meetings are of necessity short and general, more a sizing-up than anything else. Don't exaggerate or belittle your project, or feel you must act definitely about anything. What you need at this point is a sense of whether you can get your house built when you want, so you can find out when to buy a lot. If three builders each say they'll be available next year for a project like yours, your schedule is pretty safe. A builder may now steer you toward an architect, plan books, or some specific house designs he is pushing. (I'll describe hiring builder and architect later in this chapter.)

In the spring you should start shopping for lots, talking to architects, and refining your ideas about location, design, who to hire, and so forth. Now is a good time to ask about financing at your bank and meet with your accountant or financial adviser. Get your budget figured out well ahead of project-estimating meetings. By late spring you'll probably be deciding on hiring the architect and builder. This should be settled by August and confirmed with all concerned. You should own your lot by then, or have a solid purchase-and-sale agreement for one.

You'll have the fall to ratify the drawings and specifications, and shop for hardware, cabinets, and chimney styles. Your architect or builder or you can run the gauntlet of government: percolation tests and planning boards and zoning and wetlands acts. Do this in the fall. Perc and soil tests are often the bugaboo, because in some towns they can be done only in certain seasons. It may be possible to put in your foundation now to get a jump on muddy spring, although setting it in cement ties you in to a particular design.

The winter of 1990–91 will find you approving final drawings, arranging financing, and visiting your lot to consider siting options. Stay in touch with the builder to check on scheduling. When spring comes, everything's in place for an early start. (In building areas without the onus of frost, this timetable can be reproduced in other times of year.) You'll be moving in when others are crying the School's A-Startin' and the House She Ain't Done Blues.

Choose a lot

Usually, your first step in having a house built is finding a place to put it. There's lots of advice available on this subject, as with any profitable business. Location matters, schools matter, commuting distance matters, real estate values matter. What may matter most of all is whether your prospective lot is zoned for residential or some other use. Check with the planning department of your municipality to find out. Development comes very quickly when it comes, and your rural hideaway could get a chicken ranch for a neighbor any time.

As always, first figure out what you want to have when you're done. Do you want a small lawn that's easy to mow or a truck farm? How many buildings are you putting up? Do you need lots of flat, open land on which to spread out? Do you favor seclusion or a tight neighborhood? Do you want a garden, a great view, jogging trails, a home office, or room to house a collection of antique sleighs? Write it all down, in priority order if you like. If you're lucky, the list will keep you from being too irrational when you choose a lot the way most do: because you like it.

A rule of thumb is that the lot should cost 20 to 25 percent of what the house costs. Rules are made to be broken, but the notion of relative value makes sense. You won't want to cram your undersized lot with the mansion you built with the money you saved on the lot. And you won't want a sorry-looking house in a precious new development. The lot choice and the house choice must be made in tandem.

Try to consider all your costs when comparing the values of various lots. Buying a cheap lot and changing it to fit your plans may, in the end, be as expensive as buying one that is more suitable to start with. Putting a long driveway through a wet area is a good example. Clearing vegetation, grading, and landscaping can dramatically increase your land costs. A landscape architect or excavating contractor can help estimate the expense; pay one a consultation fee for an hour inspection before you buy. Then you will be able to compare the true values of different lots.

Even if you have only a rough idea of your house plan, consider it as you walk around the prospective lot. There's plenty to think about, so try to visualize your house there. You and your guests need room to drive in, park, and turn around. Should the dining room face the view, or might you use it only after dark? The sun should brighten the morning kitchen. This maple should shade the deck. You can't put the garage near that wet spot, but can you dig a small pond there? There may be many options for each area. If you can't decide what to do, or if a problem seems insurmountable, get some help.

Few people hire a landscape architect at the beginning of a job. House architects often recommend them, though, and for good reason. Very important to architects is the approach and first view of the site and buildings. And it's only right that a first-class house be sited and landscaped to show and use it to its best advantage. Working with a landscape architect, you can produce a site plan that makes your lot do all it can for you.

Most people want to spend their limited resources on the house proper. They can and want to do the "yard work," but put it off until the next summer. I'm not saying you must do all the landscaping, grading, or planting right away. What I do recommend is planning ahead. This way, you won't put in the garage, which blocks a backhoe from reaching the back yard, before you install the pool. An overview of the whole lot, with proposed uses and access for each, is a great asset to you and everyone who helps you plan and build your house.

Siting of the house is often suggested, sometimes dictated, by the peculiarities of the lot. And siting is always regulated by local zoning ordinances, which you can get from your municipality. Generally, these say how close to your lot lines, wet areas, and streets you can build. They say how far your well and septic system must be from each other and from naturally occurring water. Zoning laws may restrict the use to which you put your house; for example, sometimes a dentist can practice from home, sometimes not. Maybe you can do the site design yourself if your lot is reasonably uniform and you don't have grandiose plans for reworking it. Again, any time you get swamped, hire some help. You may even find a local landscape architecture school that will take on your problem as a project.

In the last few years, much has been made of solar orientation of houses. You should definitely plan your house and its siting to take advantage of the sun. My only caution is that you not become so dedicated to a certain orientation that you disregard your goals. If the best views are northeast, then by golly put a big window on that side and cope with the cold somehow. Principles are useful, dogma is not. Keep your mind open, even after you've decided on a particular design. Always remember, the cheapest thing to change is the plans.

Perhaps the most difficult thing to decide is the layout of the driveway. In northern climates, if driveways could be built in the winter, they would probably look different. Commonly, driveways take a straight line from the garage to the street. In winter, a sharp turn may mean you won't make it up the driveway. The typical driveway leaves little room for snow to be piled by plows, so the homeowner and plow driver are frequently at odds. A narrow driveway snaking through the spruces may look nice but it's hell to plow.

If you're planning the drive yourself, lay it out, decide how wide it will be, then make it half again wider. If the drive is long, you may want a section where two cars can get by each other. Household traffic usually comes from one direction on the street; widen the entrance on that side so the turn-in will be easier. Unless your house is quite close to the street, you'll want space to turn around in the drive. This takes a lot of room, especially for delivery trucks. For the same reason, curves should be gentle. Remember, when you plan the driveway, that most everyone who uses it will tend to take the easiest path along it. Keep things like stone walls and fancy bushes away from bends and parking turnarounds. An intricate and attractive drive will give you pleasure only if you can get in and out in February.

The lot you buy should already have passed a percolation test if your area has no municipal sewer system. This test is a simple matter of digging a hole in the ground, pouring in a measure of water, and timing its disappearance. Many towns require a deep pit, sometimes eight or ten feet. An engineer conducts the test, and a backhoe is usually hired for the digging. Get a copy of the test report. It will contain a drawing of the soil profile, showing soil types, the presence of ledge and ground water, and other features for the depth of the hole. Though conditions vary from place to place on the lot, chances are your foundation-area soil conditions will be like those in the test pit. This is vital information for your architect and builder as they plan your house.

Codes will likely dictate how far from the test pit the actual septic system can go, usually just a few feet. If the test on your lot was not done near where you'd like to site your house, or no test was done at all, you have problems. Check with the local building official or inspector. If you don't like the location of the perc test and want to do another, he'll be able to tell you what times of year it can be performed. A small matter like this can alter your whole building schedule. I recommend that you not buy any lot that hasn't been "perc'ed" and passed.

Hire a builder

The important thing you are buying when you have a house built is, of course, the house. The fun or furor of having it built will fade before the paint does. Though I will discuss getting along with crews and chiefs, what counts is what they produce. It is to this goal, I believe, you should apply the most thought. Choosing a builder is the most important decision you'll make in this undertaking.

You should hire that builder first, before an excavator, surveyor, or architect. He has his hands on every part of the house, and knows best what's right in each area. He understands exactly how such-and-such an insulation baffle in this bay window will preserve the ventilating air flow. He framed the partitions, and knows how and where to fasten the cabinets. An experienced custom builder has coaxed the best results out of countless subcontracts. He has arranged deliveries of materials and sent back the junk. He has run his business successfully, or he wouldn't *be* experienced. Someone who can do all that well can build your house well. He is invaluable to you and your architect in suggesting solutions to design problems. Hire him first, because he's the one with the broad overview of house building you need.

This does present you with a dilemma. How can you put the project out to bid if you've already got a builder? Won't the free-market virtues of competitive bidding keep money in your pocket? There's no right answer here. Either you get a good builder, loyal because you chose him out front, or you get the bidder you choose. If you try to have it both ways — send for bids after hiring a builder — everyone will be justifiably angry.

As a builder, I value being hired right at the beginning. I'm all for competition, but I like trust more.

What might follow here is a consumer's guide to choosing a builder. Consumer hints are useful when you buy some goods and services, but a house is neither of these. One way of losing a good builder is to apply to him magazine-article techniques for ferreting out the con man. Respect is required on both sides of the checkbook.

The attribute you want most in your builder is competence, best viewed in the finished product, the house. (In later chapters, I'll give you guidelines for judging his competence by the work he's done. It may be hard to separate your perception of the house you're looking at from that of its furnishings and maintenance, but try.) Judgments of workmanship should make up about 75 percent of your who's-the-builder decision.

Fifteen percent more comes in his contracting ability — making the job run smoothly, accommodating changes while sticking to the schedule, and helping the subcontractors achieve their best performance. These qualities you can only guess at from others' experiences — general reputation and recommendations. If you've lived in the area and have used tradesmen (plumbers and electricians, for instance), find out what they think about so-and-so. A lumber yard might suggest someone, perhaps more loyally than subjectively. Also fabulously subjective are your friends and colleagues. Besides any hidden agendas in their counsel, personalities play a huge role in house building. In the end, fortified with recommendations galore, you'll still be guessing how this fellow will work out for you.

Ask the builder if you can see his crew in action. Get a range of times when you can visit, not a specific appointment, so you're not treated like royalty. If you visit a job and find almost everyone working, the place fairly well picked up, and the building materials stacked in some kind of order, that's a good sign. An after-hours visit is out unless you check first with builder and owner — they're both liable for accidents on the site.

The last 10 percent of your decision should be based on a combination of cost, communication, and sympathy. The cost is the builder's bid, a figure that means, "I will build this specific house on this specific site for this amount of money." If he doesn't put much thought into his bid price, he's crazy. If he does, he's including all his costs for the work, a certain amount to cover contingencies, and a fair profit for the risk of making it all come out right. (See the next chapter for help in deciphering the bids you get.)

Communication covers all the business between the parties. Everyone involved in your project won't be equally articulate. But the more everyone tries to keep everyone else informed, the better. The plans, specs, contract, change orders, and verbal agreements must be understood by all concerned. This means writing everything down. Writing is work. It costs money that can't be spent on the building, and no one really wants to spend that money. But it is insurance, pure and simple, that you'll get what you want. You must insist on it.

Sympathy means respect. It is not important that you learn everyone's kids' names or the jargon of the trade. But to be treated well by your crews, you must acknowledge their competence and style. You screened them for suitability to your job; now leave them alone when they work. Being interested in what they're doing helps, and informs you about your house. But visit the site near the end of the workday, or, if you can't, in the late morning. This is when jobs seem to run the smoothest, and the foremen will have the fewest supervising duties. Adjusting your schedule to someone else's is respectful, and interrupting someone's work is not.

Sympathy does not necessarily mean simpatico. A careful, conservative builder will likely build your newfangled palace of dreams better than an untutored dreamer from the same church. It does matter how you feel about your builder, but much less than it matters how you feel about his work. A project with towers, flying buttresses, and 121 different windows needs a stodgy builder more than does a three-bedroom, bath-and-a-half Cape. If you want care, then you must buy care.

This is fundamental: *If you don't like any of the characters you come across, or they don't help you the way you want, move on right away*. The beginning is the only easy and fair time to drop them. It's worth the time it takes at the start to find the right people. The work that lies ahead is demanding and requires close cooperation.

One way to find a builder is to drive around, looking for houses you like. In your travels, take note of, say, five relatively new houses that you are drawn to and try to find out from the owners who built them. The local Registry of Deeds will provide the current owners' names so you can call first. You may have to ring doorbells. Most people with houses they are proud of shouldn't mind a small intrusion, but judge for yourself. Who knows? Some might even offer you a tour. What you're after is a list of competent builders.

In the meantime you will doubtless talk to others who have bought houses or had them built. Solicit their opinions of the builder, and the architect if they used one. Ask if you can tour these folks' houses. A tour involves starting in the cellar and proceeding up through the house, looking at every part of the building and mechanical systems, windows, doors, trim, drywall, cabinets, plumbing, floors, and on and on. On any inspection, keep your eyes open and your ears half closed, so you can form your own impressions of the building. Don't rush these initial explorations; this whole step may take months, so give yourself the time. What you're learning is how to look at houses and how much you don't know. And you're adding to your list of builders.

Even if you think you're incapable of judging quality in houses, after looking at a few with that in mind you'll improve. You'll start seeing gaps between adjacent materials — misaligned trim and lousy drywall. A builder's forethought will beam out at you from orderly cellar ceilings. Stains on the siding will show where the ice backs up under the eaves and runs down the wall. You will learn a lot by purposefully inspecting six or eight houses.

Look over your list. If because of your inspection or an owner's diatribe you have doubts about any builder, shuck him. One or two prospects should stand out. Ideally, now you're ready to inspect other examples of those builders' work. But by this time you might be toured out, or the prospects for finding other houses to look at might be few. Save your strength for those recommended by the builders you call.

Now call the builder you think most likely to give you what you want, then the next, and the next. This might be in the spring, perhaps April or May of 1990. Tell him you want to have a new house built the following year, you're trying to get an early start, and what does he think of architects? The answer will tell you a little about the builder. A short, rational explanation of the pros and cons of working with architects is a good sign. So is agreement with the idea, or his saying, "It depends on which one." If you intend to use an architect, you'll certainly need a builder who can work with him. If you don't, you'll still want an open-minded guy to build your house.

The first call is easy. When you reach someone, tell him who you are, how you got his name, and, in twenty-five words or less, what kind of project you have in mind. Tell him roughly how big and where it will be, and add a few descriptive passages. Skip dollar figures, but tell him what your preferred timetable is. (Remember, this call is nine to twelve months before you want to start building.) Don't flatter him; you're not trying to curry favor, just find a builder.

By the time you make these calls, you'll have an idea which builders you'd like to meet face to face. Be patient trying to set up a meeting; probably everybody is busier than he'd like. A fellow who misses a meeting without calling to reschedule it, though, should be passed over. That may be harsh, but you are trying to weed out the gross problems first. And scheduling work and appointments and keeping them is a big part of the builder's job, and of his usefulness to you. If this eliminates your whole list, run through it again but try to find someone else, too.

In their initial meeting with a builder most people ask about square-foot prices. At this stage, a prudent builder will fill his response with qualifiers like, "Normally, my houses are around sixty-eight dollars a square foot," or "This house should cost between eighty and ninety dollars a foot," or the encompassing, "It all depends." Very few customers come to a first meeting with the completed plans for their house; they simply aren't ready. So how can the builder give you a price? You are asking for a custom-built house, ordered the way you want it, and you don't even know yet what it will look like! You should expect much hedging from the builder, and suspect a glib answer to your square-foot-price question. I usually come up with a range broad enough to include almost anyone's budget.

If the builder can show you a recently completed house something like yours and tell you how much it cost, that is more relevant. But no matter what, your house is not that house and won't cost the same amount. I know you're trying to eliminate from consideration those builders who are grossly out of your spending range. But a square-foot price is inadequate to that task, too. Don't get hung up on any preliminary price, simply because it will change so much before you sign anything.

Another problem with off-the-cuff pricing is that the first figure you hear is the one you'll remember best. You will compare any revised price to that one, no matter how invalid it was. Imagine a situation in which you received two preliminary estimates for a bathroom remodeling job, $6,100 and $7,800. You asked both builders to give you final prices after the details were worked out, and they were identical — say, $7,300. Other factors being equal, would you hire the $6,100 man or the $7,800 fellow? (There is no correct answer here.) You still judge these guys by their original prices, even after you throw those figures out the window.

The prices you get for your project are apt to be higher than you can afford. They're always higher than you hope, and two things are bound to happen. First, you'll be shocked. Your frail but visceral dreams will get elbowed in the ribs, and it will take time to recover. The surprise might have come because you conspired in a design too rich for your budget. It's natural to want the most for your buck, and push the limits of your budget with an ambitious design. But scaling back your plans when the bids come in is very hard, and often means significantly compromising the design.

So you are presented with a conundrum. You can't get a definite price on a building until you have specified the building. You can't write specifications without a design. You can't produce a design without an idea of how much that design will cost to build. The architect may or may not be good at predicting the "built price" of his design, but you'll have to be guided by his judgment. In any case, you should shoot for a design he would price somewhat below your budget limit — say, by 5 or 10 percent — to avoid having to make major changes later.

After you overcome the shock, the second thing that will happen is that you'll later put back a lot of the stuff you just took out, coming up with the extra money somewhere. This occurs on lots of jobs, enough so it just seems like the natural order of things. Partly responsible is the pressure not to change a design that costs too much but is virtually finished, after all. One customer I know put back the whole second floor she'd earlier abandoned! Be warned, though, to save some of your 3-to-6-percent money for the rest of its original uses.

An experienced builder will have seen this phenomenon before. Don't be surprised if he doesn't get as agitated as you do about the early chopping and slashing — he's used to people in shock. Since you hired the builder first, he's now available to help you and your architect see which cuts will save what dollars, and which ones won't show up too much in the finished house. All this is a trying time best avoided, but you can make it through. The builder wants this project to move ahead, and will likely help you any way you need it.

The questions you're likely to elicit at your first meeting with a builder are: do you own a lot, and do you have an architect? The first one's easy, the second informative. If you have your lot, or are about to close on one, the builder knows you are serious. If you don't, the builder will probably set you back on his mental list of prospects until you have a deed in hand. You are right to call him before you own a lot, but expect polite interest, not enthusiasm.

Since you don't (or shouldn't) have an architect at this point, and may not be sure whether to hire one, the second question is important. If you're pretty certain you want to hire an architect, the builder can happily retain his connection with you without having to produce anything yet. Good builders are usually busy, and they don't mind being taken off the time hook. It's still worth something to you, ceremonially and practically, to have contacted builders first.

Both members of a customer couple should come to every meeting. This may be hard, since there may be several with different builders. But it keeps everyone equally informed. The decision of which builder to hire is the big one, so share the power and the responsibility by sharing the education.

If you have set up meetings with different builders, you should try to tell each one the same things, show him the same materials, and use the same candor each time. Because each builder is apt to teach you things you hadn't thought of before, it's hard to give the next one the same information you gave the previous one. Since you have no plans yet, the builder will be trying to guess from your presentation just what kind of house you have in mind. For your own benefit, get a true comparison by treating each case identically. Later, if you get to the point of sending out for bids, uniformity is mandatory.

You want to get him to talk about your house. You should be discussing its design — who's going to provide that? — even though the specifics are still a minor matter. You should talk about scheduling — how much lead time does this builder want, and how long will a typical project take? Remember, you are only seeking a builder now, not hiring one. This first meeting needn't last long.

Ask each fellow if he'll be on the job site every day. Just as the architect is the crucial link between your imagination and the blueprints, the builder shapes the planning into a completed project. Even if you have no architect, and the plans you bought are rock-solid, you need quality control. This is the builder's job, and he can't do it if he's at another work site.

For a small example of this, take the placement of electrical boxes. Suppose in the family room there is a thermostat, three light switches, and an outlet below, crowded between a door and where a hutch will go. The height of the boxes won't be a problem, but how about those three switches? Are they ganged side by side or stacked one over one to keep them from being hidden behind the hutch? Does the builder even care?

How many rooms have you seen where a switch was jammed against the door casing, or stuck behind the back of a door, or plopped in the middle of the wall where the family portraits must be arranged around it? You need an aware and informed builder on the site to champion your interests. Find out at your first meeting how the job will be run and who will manage the field crew. The builder himself is often best at this job because the building stands over his signature. If there is an insulating layer (a foreman) between you and the builder (the guy you pay), you may have some trouble getting what you want.

In the first meeting you should expect the builder to offer you a combination of impartial advice and moderate salesmanship. It's his job to find out whether he's the right fellow to take on your project. He may suggest changes in your approach, budget, or expectations. He may air the advantages of hiring him over the next guy. You don't want to be deafened by his horn tooting — you're there to learn. You should come away with some new ideas about your project and about the builder's world as well.

You've sounded out the builder on architects, in general and in particular. Trot out your notes on houses you like, and your magazine pictures and napkin sketches. This is not a formal presentation, and it's worth only a few minutes. What you're after is whether the builder thinks he can do what you want, when you want. Find out if he has built anything similar to the stuff you like. Ask him if you can tour two or three examples of his work he thinks will help you decide to choose him.

The builder may want to guide your tours, or may release you to the hands of his former customers. Somehow arrange to talk to them without the builder present so you'll learn more. Most people carry strong memories of their house building for years, and might let you in on them. What you'll hear will be quite subjective. Only by sampling several tales can you draw a considered conclusion. You'll find your background in looking at houses very useful on these walk-arounds. Remember, too, that these are the houses the builder wanted you to see, his best combination of customer and construction.

Who designs the house?

Use these early meetings to decide where to go for your house plans. What your builder will want from you, usually right off the bat, is a clue to how the project will be designed. Several possibilities exist. The simplest is for you to choose from several standard designs the builder may offer from his repertoire. The result won't be a custom-built house, but it may suit you. The plans will be familiar to the builder, no architect is involved, and few surprises will surface. If you're having a house built for you, this is likely the cheapest way.

Close in price are plan books, where the cost of the design is spread over many purchases, saving you about 8 percent of your house dollar. The trouble with these first two methods is they rarely give you just what you want. And the chances of your house's fitting graciously on your lot are poor. People often tell me they look at many plan books and hundreds of designs and don't find one that fits them. The quality and completeness of these plans vary considerably. Several hundred dollars is too much for plans that "sort of" fit and are incomplete and difficult to read.

The next possibility is the builder drawing his own plans to your order. This is still the way many rural projects are designed, and has definite merit. The best feature is that the builder is in on the design from the beginning. This helps him advise you on practical solutions to design problems. Speaking directly to your builder, without an architect as translator, you should be able to build just what you want. The stumbling block of this choice is the limited design talent of the builder. If what you request is beyond his skill, you won't get it.

If at the first meeting the builder suggests that he design your house, here's what he might be offering. He can work with you to sketch a design and layout. He may then hire a draftsperson to make drawings. He will establish a specification list, and show you catalogs of hardware and windows, toilets and trim. He will ask you to approve all the steps he takes. This whole procedure will cost you somewhat less than an architect's services. You may like the result, or you may not. It's a little more down-home than having a separate architect. The simplicity of fewer minds working together sometimes helps. It's more work for the builder, and he can't always take it on. And some builders offer this service and some don't.

There's another factor at work here. If you want a suitable house and you don't hire someone trained in house design, you'll have to get quite involved yourself. Design/build outfits and traditional architects do require input from you, but you can easily depend on them to lend an expert touch to your project.

On the other hand, you can spend one hundred hours or more assisting and advising on a nonprofessional's plan. You may consider this fun and challenging, or beyond your interest.

In a design/build organization you can in theory have the best of both worlds. You should get the "checks and balances" of a separate designer and builder. You should get individuals who have gotten good at one thing, not spread themselves thin learning two. It is also possible that you will get one principal you like and one you don't, or a company whose management is in conflict about its goals. The close affiliation of designer and builder may mean built-in limits on the freedom of either. Viewing completed houses and talking with their owners will give you clues to the firm's performance. Discussing your project's design possibilities at a conference will help you determine competence and compatibility.

Someone has to design your house. If your taste runs to the unusual, fancy, austere, sophisticated, dead simple, or highly detailed, you should get an architect. The vast middle ground between design highs and lows is the builder's territory. So tell him at this first meeting that you're not sure — if you're not — about needing an architect. If the builder you're talking to thinks you need one, ask him to recommend two or three. Custom builders often have some experience with local architects and have found a few they can work with. Compatible with him may not translate into compatible with you, but it's a start. You can choose your architect much the way you found your builder. Integrate tour results and recommendations into a list of prospects.

The traditional client-architect-builder triangle has a lot to recommend it. Each party has separate and fairly well defined responsibilities. In the successful triangle, the client pays the bills and has final control over the design; the architect translates the client's wishes into orders for work while following his own inspiration; and the builder produces the real thing. All the roles overlap somewhat, yet each is vital.

It would be great if you could move automatically from the feelings you have about houses to a plan. This plan would reflect everything you desired, wrapped in a stunning yet economical package. You could take the package to a builder and a few months later move in. The builder would like this, too, as he would know he was building just what you wanted. Design problems would be history.

Absent such a grand scheme, you will probably try to sketch some of what you're after yourself. Many sheets of graph paper later, you'll acknowledge the difficulty of integrating a working floor plan with the exterior and details you like, and you'll go shopping for help. Here's what you might find.

Architects are professionals licensed by their state. They must pass 4 years of college plus 3½ of grad school to get a Master of Architecture degree. Three more years of apprenticeship to a licensed architect earns them the right to take the licensing exam, and 22 percent pass. Another avenue of training is a 5-year undergraduate course that results in a Bachelor of Architecture degree. Five subsequent years of apprenticeship qualifies one for the exam.

Architecture schools end up aiming their students in certain directions. Graduates trained primarily to design houses are rare, because the bread and butter comes from projects for business, industry, and institutions. Obviously, a specialist in houses is a good choice for you; you'll get the benefit of his wide experience. As with choosing a builder, try to look at an architect's past projects. You should really like his designs, not simply be impressed by them. Collect some referrals and talk to a few architects before selecting one. This relationship will be intense for a year or so, so don't choose a blind date out of the yellow pages.

You have a few things to do before you can get down to work with your architect. If you own your lot, you can get a topographical survey of the house site. For a few hundred dollars, a surveyor will draw the lot lines and the elevation of geographical features. If you haven't bought a lot, or have the choice of several in an area you like, the architect will probably want to help you choose among them. Bring to your first working meeting your collection of pictures and notes of things you like. Know your ambitions for space, timing, and budget.

After a few meetings, your architect will develop a schematic plan of a house he thinks fits you. You might find it hard to envision a completed house from these drawings, and you can ask him to produce a scale model once you tentatively approve the drawings. This stage is the basis for the eventual design and construction, so make sure you understand what's presented to you. If you don't, slow things down *at any step of the way* until the architect can clarify matters. You may care more about where the refrigerator will go than about the south elevation, but try to absorb the whole design before you approve it.

Next, the architect will translate your schematic into drawings, also called blueprints, or plans. (If you've hired your builder, see that he's in touch with the architect during this phase. Architects usually get very little training in frame construction, and your builder can fill in the blanks.) He'll produce site plans and floor plans, elevations and sections. He will start writing specifications, a listing of roofing, doorknobs, and bathtubs. As you approve the plans, the architect will fill in the interior details, like kitchen layout, built-ins, and stairways. Further developments include plumbing and electrical layouts, foundation plans, and maybe heating schematics. He'll list windows, doors, and each room's finished surfaces. All architects don't produce all this paper; you can discuss with yours how much you think you'll need.

The drawings and specs are now combined to form the construction documents. These bind the blueprints to the builder. If you haven't hired your builder already, they constitute the bid package, and can be sent out to the three or four builders you are considering. In either case, they are the basis of the contract price. Each piece of paper in the construction documents should be initialed and dated. Plans are fluid, constantly evolving, and it's imperative that each party knows the status of the documents at bidding time. Debates can still arise because people's perceptions vary, but this is everyone's best shot.

Once the bids are in and you've chosen a builder, either he or your architect will write a contract covering the business arrangements. You can retain the architect to administer the contract. In theory, your builder will honestly and ably produce just what is drawn and specified. In practice, his view of the latitude or scope of the project may be at odds with yours. His judgment calls on the details can vary from those of the architect. This is a very sticky area. It affects many jobs, even when you've carefully

..sen the ideal builder and placed your faith in him. The best thing you can do is stay involved.

As the job proceeds, you can expect the architect to visit the site regularly and apprise you of his view of progress. That's fine, but if he thinks the builder is shaving the specs to his advantage, he should let you know immediately. Don't make the architect handle quarrels over interpreting the contract himself. You risk your house's becoming the field where builders and architects roll out their cannons of classic acrimony. There are usually enough issues right at hand without bringing in old enmities. If you're there, you can help keep the dispute manageable. The contract is between you and the builder, so settle things yourselves.

In a booklet titled *You and Your Architect*, put out by the American Institute of Architects (AIA), is a list of common questions and answers under the heading "Selecting the Architect." The last one is this: "Q: Some say that I should select a builder or contractor before selecting an architect. When is that good advice? A: It works best to select an architect first; then you will have help in understanding how to make the builder or contractor an effective member of the building team."

Does that mean builders need help to become effective? The condescending attitude in the answer is a big reason builders don't get along with architects. There are others: architects are white collar, and builders blue to gray; architects' work (the creative component, anyway) doesn't seem real to some hammer-and-nail men, yet they get a similar profit from the contract; most builders are pragmatists and see architects as flighty idealists; architects aren't trained much in techniques of house construction, yet appear, on jobs and in the authority of the paperwork, to admonish builders.

Dissension arises when the architect instructs the builder on the techniques necessary to interpret his design. Technique is the builder's turf, sometimes assiduously guarded. The architect intends no trespass, just wishes to ensure his results, so he treads a fine line here. The builder demands to know how such mischief could possibly be constructed, yet mistrusts the solution. Both parties carry the freight of any past difficulties. If the architect harbors any insecurity about his position, or the builder about his, the interactions can become volatile.

Insecurity is not uncommon. Also in the AIA pamphlet *You and Your Architect* is a useful summary of what you should know as a client. Builders get a few lines, appearing mainly as incidental characters who need supervision. The cover letter, entitled "Instructions to the Architect," doesn't even mention builders. Worse than that, it scarcely refers to the actual construction. Most paragraphs address nurturing relationships, getting hired and paid, and planning ahead to avoid liability and claims. If the AIA's goal is putting up good buildings, why is this publication so skewed? The bias appears defensive, as if the profession doubts its own authority. Instead of owner, architect, and builder marching together through the project gate, the architect appears huddled by the fence, clutching the owner, fixated more on behavior than building.

This has come to pass because architects have worked to extend their province from designing to managing the whole project. If all builders were competent, architects could present them with complete plans and specifications, and expect a literal translation into sticks and bricks. If architects always designed buildable projects, predicted construction costs realistically, and didn't tread on builders' tender toes, you, the client, wouldn't have to cope with the mutual intransigence. And you may well not, depending completely on the individuals concerned.

Class antipathy, undervaluing each other's efforts, questioning each other's judgment — these themes are always flying around the job site when an architect drew the plans. The combatants have usually learned to handle the strife. You can't help much. If you exhibit respect for both the builder and the architect, there's a fair chance they can respect each other. Hire the builder to manage and build, and the architect to design and specify. Ask that the architect be available for consultations with the builder. This cautious formulation may or may not produce a successful working team, but at least your house should emerge intact.

4

Agreements

BLUEPRINTS OR DRAWINGS, specifications, and any amendments and notes make up the bid package. One more time: *every* sheet of paper should be dated and initialed by owner, architect, and builder. Any notes on blueprints or specs, written in after the pages are printed, should be dated and initialed. Most major problems in buildings come from lack of clarity about who said what, and when he said or drew it. The bid package combined with the contract itself become the contract documents. They should remain unsullied throughout the construction and at the job site at all times to resolve differences of memory. These come up on every job.

Blueprints often look complicated beyond understanding to customers. To an experienced builder they rarely say enough. Most builders would rather look at a picture of something they are to build than read the description of it. The picture should communicate as much as possible, yet not be obscured by the notes and figures that do so. That's why a complete set of house plans can run to fifteen or twenty pages, including heating, plumbing, and electrical layout. All you really must know is whether the builder is satisfied with the quality and level of detail of the drawings.

Complete drawings (as the blueprints are known) show elevations, plan views, sections, and perhaps a rendering of the entire house and certain parts of it. Elevations are drawings looking horizontally at something, drawn without perspective or seeing in through the windows. Plan views are from above, and include floor plans, foundation plans, roof plans, and those of details like closets and cabinetwork. A section is what's exposed by slicing — like a cake would be cut, down through a detail — and showing one side of the slice as an elevation. A rendering is like an artist's sketch of a building or detail, often with accoutrements like towels or shrubbery. The rendering isn't necessary for building but helps customers get a feel for the look of the thing.

Specs

The specifications are equally important. They are a list, roughly in the order of installation, of all the parts of the house, from the type of concrete in the footings to the model number of the towel bars. Perfectly complete specs are gems rare and desirable. Architects sometimes sidestep drudgery by inserting American Institute of Architects specs (generic boilerplate) throughout the spec list. This limitlessly thorough stuff regularly insults the careful tradesman and is ignored by the rascal. Good specs are a complete parts list, with infrequent incursions into methods of work.

So many buildings have been designed so similarly for so many years it would seem possible to catalog specs of every component. Pulling these out, perhaps from a computerized data base, and assembling them in your house's spec package would be a snap. Yet time and again builders run into parts of structures the specs insist should be vastly overbuilt or fussy when simpler would be better. And virtually every job's spec list leaves out many details, or makes assumptions, or skips whole sections entirely.

This isn't a big problem if your builder is working to his own high standards. If I find something in the specs that lets me get away with something I shouldn't, I try to point that out in the precontract planning. Ask, and pay, your builder to do the same for you. In an hour or two he can read the specs and note areas he thinks need further definition, or those he can't find at all.

What follows is what I consider a *minimum* spec list for a sample house. If you think your house is fancier or you are fussier than most, more detail will be required for each entry. Most useful to all parties is the name, manufacturer, model number, grade, color, size, and description of *each item*. It would be a rare house project that would need a shorter list than this.

SPECIFICATIONS, ARNOLD RESIDENCE
WELLFLEET AND BREWSTER, ARCHITECTS
TRUEROW, MASSACHUSETTS

Project Architect: Robert Brewster, AIA
Owner(s): Michael and Grace Arnold
99 Western Court
Los Angeles, California
Project Location: 10 Salt Street, Truerow

These Specifications are intended to accompany Construction Drawings by this firm, dated __/__/__, pages one through fourteen. Together they form the Bid Documents for the Arnold Residence project. Only these documents are to be used as a basis for bids. General Contractor shall perform or cause to be performed all work in accordance with these specifications and leading trade practice.
These Specifications are dated: __/__/__

By:________________________ Robert Brewster

SPECIFICATIONS

Water Supply — To be drilled well by owner, construction permits contingent thereupon, to be completed before Start of Work. Owner will supply pump, pipe and fittings to wellhead, plumbing subcontract to start from there. Pump to be submersible, two stage, 220V.

Sanitary Sewer — To be a private disposal works, as per application #3297, approved by the Truerow Board of Health 4/13/89.

Foundation — To be poured concrete, 2500# mix, 8″ thick, on 18x10″ continuous concrete footing. Footings under columns to be 2′-6x 2′-6x10″. 2/#6 reinforcing rod continuous at half height of footing.

Dampproofing — Bituminous asphalt, two coats, brush application, from outside of footing to grade. Form tie cavities to be filled completely with trowel grade patching asphalt.

Foundation drains — To be continuous around building at footing height, and consisting of 4″ dia. rigid PVC fittings and pipe sloped to daylight, pipe perforated where next to the foundation. Drains to be covered with 1″ washed stone a minimum of 6″. Drain stone to be covered with approved filter fabric.

Backfill — To be suitable coarse fill or bank run gravel. Backfill material must be free of organic matter, clays, construction materials, and rubbish. Material must permit the rapid dispersal of water under all site conditions. Backfill must extend out minimum 3′ from the foundation, and in height from drainage stone to within 6″ of finish grade.

Finish grade — To be topsoil reclaimed from site, restored minimum 8″ below the top of the foundation. After time of possible injury, finish grade to be smoothed, raked, and seeded. Additional topsoil as needed will be furnished according to the excavation allowance. Excess fill or diggings to be trucked away.

Sills — To be pressure-treated 2x6 lumber, attached with min. ½″ dia. anchor bolts in foundation, min. 6′ o.c. No section of sill shall have fewer than two anchor bolts.

Columns — To be concrete-filled steel lally columns, 4″ dia. with Springfield caps and bases, as located on foundation plan.

Girders — To be built-up, from three 2x10′s, KD [kiln-dried] Douglas Fir #2 or better, splice-joined over columns. Members spiked together 12″ o.c.

Joists — To be 2x10 KD Hem/Fir, #2 or better, 16″ o.c., to yield a uniform 50# psf total load rating. Note solid bridging over girders. Ell joists to receive wood cross-bridging at half-span.

Floor sheathing — To be ¾″ Fir plywood, underlayment/sheathing, tongue and grooved and nailed ± 8″ o.c. with 8d hot-dipped galvanized nails.

Floor insulation — To be 6½″ thick unfaced fiberglass batts with wire supports, R-19. Celotex white board ceiling below.

Wall framing — Exterior to be 2x6 KD SPF [spruce-pine-fir, a lumber grading classification] construction grade, 24″ o.c. Southwest cathedral gable to receive horizontal nailers 36″ o.c. for board siding. Interior to be 2x4 KD SPF construction grade, 16″ o.c., firestopping at 8′ lifts at cathedral walls.

Window and door headers — 2/2x10 KD Fir, with 2x3 separators and glass fiber infill.

Wall sheathing — To be ½″ CDX plywood, fastened with 6d hot-dipped galv. nails.

ulation — To be 6½″ thick fiberglass batts, R-22, with 6-mil retarder on face of studs.

g — To be one half by six vertical grade Red Cedar clapboards, 4″ exposure, except southwest face to be diagonal 1x6 T&G Cedar boards. Nailing to be 5d hot-dipped galvanized in random pattern, approx. 8″ o.c.

Roof/ceiling framing — To be all KD, 2x8 SPF rafters, 16″ o.c., 2x10 ridge board, 2x6 collar ties, 2x8 ceiling joists, engineering to calculated loads of 15# dead, 30# live. Ceiling will be firred with nom. 1x3 wood firring. Cathedral ceiling roofs will be framed with 2x12 SPF construction grade rafters, 24″ o.c.

Roof sheathing — To be ⅝″ CDX plywood, fastened with 6d hot-dipped galv. nails.

Roofing — To be Class C asphalt/fiberglass shingles, Slate Blend, 220# per square, nail application, over 15# felt, with continuous 36″ wide Bituthene (or equal) membrane strip along eaves and centered under valleys. Valleys to be 20″ wide .019 copper flashing. Chimney to be flashed with lead, counterflashing to be copper, soldered copper cricket.

Roof insulation — To be 9″ thick fiberglass batts in rafter bays, R-30, with 4-mil poly vapor retarder on rafter faces, and polystyrene vent channels full length in each rafter bay.

Ceiling insulation — To be 9″ thick chopped blown Insul-safe fiberglass, R-30, with 6-mil poly vapor retarder fastened to firring.

Chimney — To be concrete chimney block to roofline, Heritage Mixed brick above, 8x8″ fired clay tile liner with furnace cement at joints. 8″ thimble, 6′-4 center above slab height on southeast side in cellar. Cleanout door on southwest side.

Roof/attic ventilation — To be continuous black aluminum ridge vent by Air-Vent on all ridge areas, and continuous 2″ white Air-Vent soffit strip on all horizontal soffits. Cathedral ceiling areas and all roof areas where ceiling insulation might interfere with ventilating airflow will have preformed foam ventilating channels.

Windows — To be Pella Clad awnings, ventilating or fixed as marked on drawings, and Pella Clad casements or fixed units, as listed in window schedule. All windows to be white exterior clad, with full screens, white grills, Low-E glass, extension jambs as nec.

Doors — Front to be Peachtree Avanti, inswing, 3′-0x6′-8, with Schlage A-series knob latches Plymouth style and single acting deadbolts. Rear is Peachtree Atrium door, 6′-0x6′-8, right active, same hardware, keyed alike.

Outside pavers — To be brick Town Hall pavers laid in sand, running bond.

Exterior finish — To be oil-base primer plus two coats oil-base paint on trim and doors, Benjamin Moore CWF on siding, two coats, second coat after six months, spray-plus-brush-out application.

Interior walls and ceilings — To be 5/8" thick drywall, screw fastened direct to framing except flat ceilings to be firred. Acoustical insulation will be provided in living room and dining room partitions as indicated by 3½" fiberglass batts. Family room ceiling to be knotty pine 1x6 matched boards, clear finish.

Interior doors — To be sized as per door schedule, all 6′-8 high, birch, clear finish, hung on solid wood jambs, three dull brass finish butts per door, passage hardware F-series by Schlage, Plymouth style.

Door and window trim — To be 1x3 clear pine, sanded and nails filled, 1/8" roundover on outside edges, natural finish.

Base — To be 1x6 clear pine as above in all rooms, 1/8" roundover on outside edge, except tile cove to match floor tile in master bath.

Closets — To be finished as required by the drawings with ¾" plywood shelving with pine facings, rods and rod supports where req. Storage and linen closets will be assumed to have five full-width shelves as above, 20" deep.

Cabinets and vanities — To be chosen by owners, installed by contractor, allowance in contract.

Countertops — To be plastic laminate over HD particle board, site built, as per drawings.

Vanity countertops — Same as above.

Staircase — 13 risers @ 7⅜" ± 1/8", as drawn. Treads to be 1 1/16" x 11" Clear Red Oak with return nosings where applicable, and oak scotia under all nosings, starting step Morgan M-780, starting newel Morgan M-765, starting volute Morgan M-719, hall newel Morgan M-885, handrail Morgan M-720, balusters Morgan M-891. All staircase joints to be glued. Risers, skirting, newels, and balusters to be painted as typ. interior trim. Treads and nosing to be stained and three coats polyurethane applied. Handrail to receive varnish-stain finish.

Cellar stairs — To be rough stringers, Yellow pine treads and #2 Pine risers, paint finish, no wall below or beside in cellar.

Flooring — Downstairs to be Red Oak strip, #2 and better, sanded smooth and with three coats polyurethane finish. Upstairs and Powder Room to be carpet, vinyl or tile, as indicated on drawings and room finish schedule, and on allowance, to be chosen by owner and installed by general contractor. Carpet allowance is $25 per square yard for carpet and pad installed. Vinyl allowance is $22 per square yard for the product installed. Vinyl and tile to receive underlayment as necessary to make surface flush with wood floors. Carpet and pad to be chosen to finish ½″ higher than adjacent wood floors. Underlayment to be PTS fir plywood under tile and vinyl.

Paint — To be two coats latex flat over drywall sealer on walls and ceilings, colors by Owner (limit one per room, four total), smooth roller finish. Trim will be clear finish; one coat sanding sealer plus two coats spar varnish.

Plumbing — To be completed as shown on drawings, with Type M copper supply and CPVC DWV runs as per Uniform Plumbing Code. American Standard Plaza water closets #2016.019, both baths. American Standard Plaza pedestal lavatories #0184.038 upstairs, Delta faucets. American Standard Spectra bathtub #2605, with Delta Scald-Gard faucet. 4¼″ sq. white Florida tile walls in thinset over Wonderboard, white grout. Efron shower door Regency #350, mill finish. American Standard Rondalyn sink @ Powder Room, Delta faucets. All fixtures to be white. Elkay kitchen sink, #ILGR-4322-R, with Elkay #LK-4301 faucet and retractable spray. Shutoffs at each WC and bowl. A washer/dryer shutoff connection to be provided in the cellar. Utility sink to be Swan #MF-1W. Two outside frostproof faucets. Plumber will provide X-Trol and pump controls, and water supply fittings as required from wellhead in, to maintain 30–50 PSI water pressure to house faucets.

Electrical — To be a 200 amp. underground service, outlets as per code, two outside outlets, switch controlled lighting in all rooms to code, two doorbell locations, lighting in cellar, ventilation fans in both baths, connections to HVAC unit, range hood, clothes dryer in cellar, water heater. Fixtures by Owner, price under allowances in contract. Range hood, fans and/or fan/lights, door chimes, yard and post lights are considered fixtures, and come under fixture allowance.

Telephone — To be based on 1 incoming line, phone jacks in 4 locations, Owner's choice, indoors. All permanent phone wiring above first floor level to be prewired and concealed. Phone equipment by Owner.

Security — Not in contract.

HVAC — To be designed by HVAC contractor as per code, based on 6300 degree days and 0° F. design temperature for heating and 86° F. dry bulb design temperature for cooling, to provide a 72° F. to 78° F. indoor temperature year-round and complete ventilation of all rooms. Furnace to be oil-fired, with one 275-gal. tank in the cellar.

Miscellaneous — Bath Accessories to be Hall-Mack as detailed in drawings, p. 12, elevation 12/4, furnish solid blocking in walls for attachment. Window shutters on southwest side only, Bird vinyl, black, to match window heights. Guttering to be white aluminum, continuous, on fascia brackets. Downspouts to lead to buried 4″ dia. ABS pipe sloped to daylight, not connected to perimeter footing drains. Interior window treatments not included.

Additions or modification to these Specifications will be made as Change Orders, discussed in the Job Contract.

When the project drawings and specifications are complete, they are packaged and sent out for bids. Of course, if you've hired your builder already, he'll be the only recipient of the package. If not, confirm with the architect's office that the bid package will be sent to the bidders you have selected. You can add one or two the architect prefers, but try not to exceed four or so in all.

Often the architect will have sent preview copies of proposed designs and specs to prospective bidders, especially to the ones you've selected. Any bid *must* refer precisely to the dated drawings and specifications in the bid package. Though later changes in the design and specs aren't uncommon, for a fair comparative bid each party must be bidding on the same project. Also, because the bidder puts considerable thought into the bid, the project bespeaks that shape to him. Changes are more cumbersome and disruptive from this point on.

Before any package goes out, call each bidder, if you haven't already, and ask if he's interested in the job. The personal contact is important; it lets the builder know *you* are interested in the hiring and the outcome. Ask the architect to be sure all bid packages are mailed or delivered on the same day. Building is a very competitive business, and no one likes to see the next guy get a head start.

Expect several weeks' wait before the bids come back. The bid package cover letter will state, "Bids must be submitted by __/__/__." Reliability is important, but bids are difficult to put together, particularly for a small firm. If you or the architect has no word at all from a bidder by the due date, that's bad. But keep in mind your goal: getting a good house built. If whoever you think is the best builder keeps you waiting, maybe you should be willing to wait. Use your discretion if the deadline passes and all bids aren't received.

When all bids are in to the architect, you'll meet with him to review them and select a builder. You and the architectural team will compare each bidder's presentation, called a proposal. The bidder should include in the proposal the anticipated starting and completion dates, the bid price, payment schedule, and terms.

Many factors influence the final bid price. The builders try to weigh all their costs accurately and allow themselves a fair profit. They consider the location of the job and the season of the year. They must ponder the national and local economic picture and divine trends in the price of materials. They'll calculate their own workload and prospects. Anything they know about you and the architect fits in somewhere. Regardless of the extrinsic judgments, the end product is a bid: I will build this house for this much money.

It's more than likely that the builder you thought best isn't the low bidder. Differences among individuals aside, it takes longer and costs more to do a good job. The biggest influence on the success of your house project is who builds it, so take a few days to make the decision. Consider the architect's opinion — they must work together, too. Mainly, fix your own verdict. Now stand up so I can hit you over the head with this again: it's *your* house and *your* responsibility, so *you* hire the builder.

Once you do, call him and tell him yourself. The architect's office will send him confirmation in writing. Often construction will be weeks or months away, so the decision may have the quality of anticlimax. You need a break from it all now, anyway. There are some other things to consider before groundbreaking, though. The contract — your business arrangement with your builder—is foremost.

The contract

A contract to build a house should be a simple thing. Frank and Mary agree to pay Bud $210,000 to build their house during 1991. Frank and Mary and Bud sign the paper, and that's that. Many houses have been built with even less paperwork. Apple Corps has built many with a one-page contract. We have come to demand more from our agreements, though, and think you should as well.

In times gone by, most of our work was for local people who were going to stay around, as were we. No party to a deal in such a setting would dare set himself up as an outcast by taking advantage of the other. Too, we often built for folks we knew. We also worked with architects, who usually demanded some kind of contract, but that was exceptional.

As the price of houses went up, the people who could afford them were more likely than ever to be professionals or professors, both at ease with the written word. Consumer awareness in the 1970s seemed to make some believe they must keep endless vigil lest they be taken. The more expensive projects often came with architects, who relied on documentation to establish their rights in the domain of owner and builder. Litigation's cloud overshadowed every manner of agreement between people. Now we won't build anything without a contract.

Contracts vary, of course. The simplest describes the principals, the work, the dates, and the price. This can be done on one page, as in the example that follows. Apple Corps uses a form like this for projects without an active architect and for smaller-than-house-size jobs. The important part of this contract, or any other, is that it refers to a particular set of plans and specifications as being part of the contract documents. At minimum, you should use an agreement like this for any building project costing more than a few hundred dollars.

CONTRACT AGREEMENT FOR CONSTRUCTING A RESIDENCE

1. Customer: Your name
Address
Phone, home and work
2. Architect: Name and phone
3. Contractor: Name, address and phone
4. Job location: Address, or subdivision #
5. Job description: (Ten to twenty words)
6. Expected date of: Start __/__/__
Completion __/__/__
7. Drawings: How many pages?__
Attached?__
Dated __/__/__

8. Specifications: How many pages?___
   Attached?___
   Dated ___/___/___

   All sheets of drawings, specifications and other appurtenant documents to this contract shall be dated and signed or initialed by the Architect to indicate they comprise the Bid Documents.

9A. Billing: Fixed price? $__________

** OR **

9B. Alternate billing: Stock-and-time? $__________
This price is an estimate only. Fill in sect. 12.

10. Payment schedule (check one):
Work completed: ___weekly, ___bi-weekly, ___monthly
Schedule attached: ___ and dated: ___/___/___

11. Change orders:
Changes in the work outlined in the Bid Documents ordered by the Owner and/or the Architect will be performed by the Contractor, at prices quoted by him. Any change over $50.00 (or $?) in value will be described in a Change Order written by the Contractor. If such changes are quoted at higher than $200.00 (or $?), written authorization from the Owner shall be required before the change is effected. Changes under $200.00 (or $?) may be approved by Owner or Architect verbally, but written authorization must follow within three (or ?) working days. Written authorization is deemed to be a completed and fully signed Change Order form for each change. Sample Change Order form attached ___, and dated ___/___/___.

12. If stock-and-time project, complete this section:
Labor rates: Supervisors _____/hr.
Carpenters _____/hr.
Laborers _____/hr.
Materials billed at builder's cost plus _____%
Subcontracts billed at builder's cost plus _____%

13. Start-up payment: How much $__________
When paid? ___/___/___
Credited to ___ First bill? ___ First & second?

14. Terms: All bills due net ___ days.
After ___ days interest at ___% per month will be assessed on overdue accounts.
Retainage of (10)% on each bill, to be paid as the final (10)% billing after work is complete.

In the event of a dispute arising over issues in this contract or in the Contract Documents, parties to the dispute will make a sincere effort to settle the matter between themselves. If this effort fails, the parties will seek assistance from the Contract Arbitration Project at Chilson Community College, Donald Cote, director. The parties agree to abide by the finding of the project.

Signatures:	dated:
Customer A	__/__/__
Customer B	__/__/__
Architect	__/__/__
Builder	__/__/__

One thing in this contract that may be unfamiliar to you is "retainage." This is an amount, usually a small percentage of each bill, held back until the whole job is finished. It is intended as insurance for the customer that the job actually will be completed, and is usually paid only when disputes over workmanship and timing are resolved.

Retainage is a normal feature of big construction jobs. A small custom house builder who doesn't get the retained amount (usually around 10 percent) of his bill may be operating your job at a negligible profit until the very end. While this leash may secure his obedience, he's not a dog, and it won't get you his dedication. He may be more worried about his running expenses than about your house. Retainage is wise for the rough and tumble of commercial projects, but not for your house.

I recommend a straightforward contract. I'd like you to stand back and see the arrangement before you in its simplest terms. You will hire me for a certain amount to build a certain thing for you at a certain time. I can't wait until the end of the job to be paid. As we proceed with the work, you will pay me in full for what I've done. My partner Rich Gougeon has often said that he'd like to just work and not send people bills; jobs run smoother without the money ax hanging overhead. If you have a contract you agreed to, stick to it.

If something dreadful happens to my customer or me, we should be financially up to date. If you want to fire me halfway through, you still owe me for what I've completed in a workmanlike manner. You may want to sue me for some reason, though no one ever has. A court may have to settle the dispute, but you *still* owe me the dollar value of the work I've done for you. Customers get themselves in much trouble by being cagey about their payments.

Suppose your builder doesn't do what he's supposed to. This rarely happens in one easily recognizable flash; more often, gradually you see you're not getting what you want, or your bills are higher than the contract price called for, or the quality of the work is below par. You're in a bind. You don't want to stop the job, but you don't like what's going on. Your power over your builder resides in emotion, money, and law.

If you entered into the contract intending to be cooperative and amiable, and you have been, you have money in the bank. Assuming that the architect and the builder are not squabbling, the materials have come on time, and the builder isn't having a midlife crisis, you should have a good relationship with the builder. Your first questions and comments should be directed to him. Not to his employees. Not to the architect. Not to your friends. If your clear goal is solving a problem in the construction, talk first to the man responsible for the construction.

Make your goal the solving of the problem, not psychological redress. If you think you're being mistreated, you may want to punish the offending party. It's fine to be emotional about the difficulties at hand, but usually you're trying to solve a physical problem. What gets your building fixed is work on the building. Your ego and angst are poor tools for construction.

The fact is, the most complete contract in the world won't get you what you want unless all concerned want that to happen. If you find you've chosen a builder whose employees don't respect him, and whose subs care little for your convenience or his, there isn't much of a way out. Holding up a payment to force some action is the commonest ploy. It works sometimes. When you get his bill, tell your builder that you don't intend to pay it until the work it covers is satisfactorily completed. Builders almost always need the money and will cooperate to get it.

Outside arbitrators or mediators can often settle disagreements by helping the parties separate the actual problems from the emotions. It's important to include in your contract some external means of settling disputes; it's good insurance and may even prevent flare-ups by its presence.

Sometimes discord gets beyond simple solutions, of course. Over the years, I've advised several people about their projects when they were contemplating suing. I often sympathized with the owners. I hate to see shoddy workmanship in any building. I especially hate to see people taken advantage of by their builder, because they must come home to their bitter pill every evening. So I think if a builder fails utterly to do what he says, won't respond to reasonable entreaties, and leaves you without a fair building for the price, sue the bastard.

The key words, though, are "for the price." If when you got the bids back you went over the whole plan to find the cheapest alternative in each area, beware. The builder has some right to expect to build roughly the house he bid on. Most house plans are changed in response to the preliminary estimate. But if you squeeze every last part of the bid to extract the profit from it, you may get what you deserve. You'll be wringing the life out of the project and the cooperation out of the builder with the same motion. If your goals, budget, and design weren't carefully aligned in the first place, you might find yourself in this unfortunate position. Plan your project to the budget in the beginning.

You probably noticed in the sample contract another, completely different arrangement for running the job. It's known in the trades as stock-and-time, time-and-materials, or (how dreadful) cost-plus. In this system, your builder submits an estimate — not a bid — reflecting his best guess of the cost of the job. He states in his proposal that he will bill you his labor at *x* rate, his supplies at *y* percent over his cost, and his subcontracts at *z* percent over his cost. Or he may bill you everything at his cost, adding a percentage to the total for overhead and profit.

The allure of this system is friendliness. The builder takes no risks on his estimate, and merely bills you for work completed. If you change plans, no paperwork is necessary and the cost is clearly on you. If you come to the job site to chat about a change you want, or about a ball game, conference time is billed at the same rate as the work it displaced. The builder and subs have no financial incentive to rush their work and can give more attention to quality. Billing is simple, dependent not on phase-completion predictions but on actual work in place or on elapsed time. The total price of the job bears a direct relationship to its cost, and the builder's "fear factor" is eliminated.

Of course, stock-and-time has its dark side. You have no guaranteed price from which to plan your budget. You'll find out where you stand every billing period, but can spend unplanned thousands in between. Disputes are harder to resolve because there is no absolute standard by which to reckon efficiency. If the builder messes something up as he goes along, you get to pay for the repair. The superb self-governing qualities of the free-market model are missing.

From the builder's point of view, you have bought with your trust some measure of control over the work in progress. If the crew takes a long coffee break, they wonder if you wonder who's paying for it. You might question your builder's business practices — immaterial to you under a contract — if you think they are costing you unnecessary cash. Stock-and-time is bolder and more personal than fixed-price contract work. Don't attempt it, though, unless you are willing to be truly involved and resilient. I often find the strains too thick an overlay to the inherent human difficulties of house building.

Customers sometimes want to work on their project to save money. Painting is often the job of choice; most people think they can paint, and some can. Good carpenters like good painters because everyone's work comes out looking its best. So you may encounter some resistance to your working on the house. Some builders think of the whole project as a unit, and your painting may not fit in with your builder's sense of the job. So figure out what you think you can do yourself, and talk about it before the contract is drawn. As long as you can work on the builder's schedule, you should be able to go ahead.

"On the builder's schedule" really means staying out of the way. If you can paint or put up shelving or even clean up the site in an area or at a time different from your hired crew, and they agree to it, then you should be OK. But recognize that a job may require territory you didn't dream of.

Say the electrician is inside, roughing-in the wiring, and the carpenters have finished clapboarding the back and are all working out front. You figure you'll get a jump on the exterior painting by priming the back wall. You check with the carpentry crew and get the universal "no problem," so you go to it. Trouble is, the electrician's four-inch hole saw comes grinding out through the back clapboards, making way for the bath fan vent. Now he must install the vent in the wet wall, or wait, or ask you to do it, or bitch to the contractor, or walk away and leave it. It's nobody's fault, really, but you've messed up the electrician's day. It costs you ultimately. Either you pay money for the electrician's trouble or you get a grumpier electrician and contractor, which gets you a worse job.

Another problem you may create if you decide to work on the job is interrupting the pattern of events the workers are used to. Any well-run job has a routine of tasks, timing, and people that is comfortable for the crew because it is predictable. Even under unfavorable circumstances like bad weather or presidential scandal, at least the workers know what to expect at coffee break and lunch time.

When the customer enters these shop-talk-and-gossip sessions, he risks straining the sustaining familiarity. Unless you really feel at home with your builders, you may have set up a slight class distinction when you employed them to do your work. I don't mean to discourage your attempts to make friends or be at ease with those you hire; we have many friends among our customers. Just keep your eyes open.

I might as well render my tool lecture now. Even if your crew agrees to your working around them, they are unlikely to extend you the courtesy of the use of their tools. Even when you can anticipate which tool you want to borrow and ask in advance, don't. The carpenter, hired by the builder, owns his own tools. The builder, not his employee, is the one supposedly profiting by his good relationship with you. But don't ask him either. It's almost impossible to know how a craftsman feels about his tools, even if you ask him and he tells you.

If you can get a builder to agree to let you work with him to learn the trade, you are both exceptional. You're paying someone serious money to work hard for you and asking him to slow down to teach you what he is doing. That puts you in a bind will probably net you less building and less teaching than bargained for. The builder is crazy to do this job any way by the hour. He has to go fast to make money and slow for you to learn. And, he'll have you officially standing around watching him all the time. How would that go where you work? It stands to reason he will make compromises to handle the pressure.

The usual trouble with this scheme comes when it's unofficial. When you spend a lot of time at the job site in order to pick up tips and techniques, you're getting something you didn't pay for. This may be OK with your builder, but it wouldn't be with me. Don't forget, many builders will swallow their words rather than upbraid or offend a customer. But the sour feelings of an inequitable relationship pervade the job. You will get less house; that is your tuition.

This doesn't mean you mustn't come around with questions and comments about the job. Just be sure you do so at a time that doesn't interrupt the progress of the work or the workers' deserved breaks during the day. Early on, you can ask when it would be a good time to show up.

Most complaints tradesmen have about their customers are about "changing their minds" and money. Money may seem the more important of the two but accounts for fewer of the day-to-day headaches, especially during a busy building season. Imagine yourself the builder for a moment.

In planning the job you have set the stage for construction. You know when the props and actors come in, and the implications of the timing of entrances and exits. The play is still on paper, but you gather commitments from the cast, peg the price of the tickets, and proceed. You can make money and escape with your sanity only if the show goes on without undue interruption. You must follow the script to get through, and herein lies the rub.

Let's say the Arnold house construction started in June. Plans were all approved, and bids from Y.O.U. Construction Company and subs were accepted by May 3. This means you broached the idea of the job with your subs in February and, after several discussions and alterations, collected their bids in ıstom builders are very alert to their customers' reactions to completed work. If people don't like something I've built, ı though they agreed to it on paper, I know I'll often have hange it. If the customer wants the change and the archi- doesn't, it takes more time. Changes entail re-estimating probably canceling and reordering some materials (suppli- don't like this); discussing the change and new price with customer, the architect, and the subcontractor; and then, of rse, doing the work. All this merely to get back to where I before the change was mentioned.

eople have to be able to change their minds, though. And n the subject wasn't discussed enough in the planning, or a influence on the decision appears (an out-of-town uncle's nion). So builders must be patient. A certain number of minor rations are never billed because that's more trouble than 're worth. Between five and ten minor changes and two or e bigger ones won't get you in much trouble on a house-size ect.

you have difficulty making up your mind, let your builder w up front, before any contract is signed. You should also e your willingness to pay for your indecisiveness. And be ing. When the time is at hand to lay out the location of the side light switches and you "feel strongly both ways," you e to know the meter is running. Builders aren't shrinks or nselors; you should acknowledge your own idiosyncrasies. If might hinder your relationship with your builder or archi- let him know somehow. This is not an encounter session, herapy, only a courtesy to those who you're hiring, and it dn't be belabored.

ır relationship with your builder's subcontractors will also ıire some thought as the job proceeds. First of all, a defini- . Subcontractors perform work covered by the main house tract. They are specialists in a single field, or a few related ds. The subcontractors you most often think of are plumbers electricians. Others are insulation installers, garage door in- lers, alarm system mechanics, tile setters, and drywall tapers. some jobs, most or all of the work done on the house is contracted. Framing, trim, wood floors, and cabinets can all

late April. The foundation sub agreed to pour the
but a wet site on a previous job pushed the date b
Y.O.U. got everyone down the line prepared for
By the time the roof was on and you were fram
partitions, the Arnolds panicked — they wouldn
the kids playing in the yard from the kitchen. T
that by moving a window a few feet and makir
ing door they could have more light in the bre
see out, too. The architect would object strong
convention until the following week. The plum
now but not for three weeks afterward. The ele
to know about the slider in order to change
panels if necessary, a special-order item long sir
the wholesaler's. The list goes on and the he
chronic. Of course, this is what the Arnolds ar
do, intercede for them and champion their inter

The whole point is, changes are a pain. They
the job's progress. They can strain relationsh
the parties. They will, they must cost you more
lent decision made in the planning stages. Wh
move to you may loom large to the builder a
many of the sub trades. In the best of circum
are an interruption in the flow of the job. At
lems they present may end up in court. You
your plans the time and attention to make t
you start.

Still, you may have a hard time visualizing
room, or wall will look like from floor plans.
architect's elevations (straight-ahead views o
buildings, without perspective) may not unvei
when the window openings or chimney or
real life is often the first time you can see w
Sometimes it's disappointing. You know
money for the things you're buying, and des
you want. And you know the builder has a
and that you agreed to have the thing built t
the plans were drawn. So you might put of
like something, hoping the finishing of it
better.

be performed under subcontracts. Apple Corps tries to do as much of the work on a house as is practical. We fit our own trim and staircases, lay and sand our own hardwood floors, and install (and sometimes construct) our own kitchen cabinets.

We do our own work for two reasons. First, like any apprehensive customer, we sometimes want the control of timing and price that do-it-yourself promises. Second, though we're slower, we can be more fussy on certain jobs than some crews we can hire. Most subs must hustle to keep their businesses alive, and some do so at the expense of their standards. Under rare circumstances I'll hire a sub based on his low bid. What I prefer, though, is to find subs whose work is top shelf, and stick with them.

In your dealings with subs, recognize that the chain of command follows the flow of money. You, the source of the cash, are commander in chief. Admiral Architect is nearby, in the service of your mission. General Contractor is next, assisted by his staff, the suppliers. The subs are his lieutenants, with their captains and sergeants deployed. Peace is best maintained by observing the hierarchy and dealing directly with the highest-ranking officer.

To wit: if you are at the site and the sub asks you for, say, the location of the garage light switch, by all means answer him. Don't, however, pressure him to install incandescent lighting under the upper kitchen cabinets when the contract calls for fluorescent, or when it doesn't specify. That decision should have been made when the specifications were written for the job. Any decision involving money means involving the general contractor.

What might happen on that kitchen lighting job is this. You come by the house one afternoon when the electrician is ready to install the lights, and the general contractor is not there. The electrician doesn't have specs for the lighting because they're missing or incomplete, so he asks you what you want. It wasn't in the specs anyway, so you say incandescent. This is the more expensive solution, but the electrician doesn't tell you that. (He usually won't discuss money matters with you because he's accountable to the general.) If the general is not on the job the next morning, the electrician may install the more expensive system, figuring on billing the extra as an addition to his contract with the general. Since the lighting has been installed by the time the general gets there, and since it's not a big expense, he may choose to overlook the additional charge, paying the electrician but not billing you for it. The electrician knows he's in slight trouble with the general, but he had to make the decision. The general is out some bucks and some hassles. You have your preferred lights, but not the full trust and respect of the tradesmen.

But you didn't do anything wrong, did you? You only answered a question, and nobody told you the answer was wrong, or that it would get you in trouble. You're right about all that, but answering the question was still the wrong move. Common sense will tell you whether your relationship with the sub allows you to question his work to his face. Almost always, though, it's better to go through the general. This harks back to the first rule: know your place. Almost any question or conversation on the job costs someone money, and unless it's clearly you, you stand a chance of bruising the contract. It's too bad this rule is necessary if it distances you from the subs, but it is the correct way to conduct the business at hand.

You can see in this the value of complete specs. When all the "what goes here?" questions have been answered ahead of time, everyone benefits from a straightforward job and a straightforward relationship.

If there are allowance, or wait-and-see, items that you must choose while the job is in progress, be sure to make your choice clear right away. If you can't decide, you might hold up the job — a Bad Thing. If you know you have a hard time making decisions, and you know one is coming, get any kind of help you need to be decisive when the time comes. It should be made clear at contract-writing time which decisions have been deferred until some point during the job, so be ready. Remember, any time part of the job is waiting on you, you are making trouble for yourself, your budget, and your house.

Keep in mind that everyone who works on your house will more or less consider it just another job. Some jobs are better than others because of personalities, working conditions, outside factors like the weather, world events, and ball games, and the concept and design of the house. You can't exert much influence except over the last category, though your personality will be felt throughout the job. I hope to make clear to you how a good job proceeds, so you can do your part in helping yours be good.

The rest of the book will take you through the actual construction of your house. It will start with you, with contract in hand, builder hired, plans and specs finalized. You own your lot. Months of planning and choosing, *your* work, is behind you. Ahead, the prospect of a new house for your efforts. There may be more information here than you care to know right now. But when you pick up your plans and specs, you'll be able to understand what you're looking at. If you take the trouble to read the rest of the book, you'll also be able to judge the houses you'll be touring when you start shopping for a builder.

The following chapters may be hard to slog through. I trust that the overview of the project you'll get here will help you set your priorities for choosing and specifying. It's likely you can't spend to the limit on each part of your house. Without a sense of the whole, you'll find it hard to see where your money's best spent. You may not want to read straight through, in which case you can use the book as a reference for specific subjects. If your specs don't mention something I've talked about, check with your architect or builder before anything is sent out for bids. While there are often various solutions to any building problem, you'll be happy to have one right answer in hand. During the planning is your last cheap chance for getting what you want. If you want to change something after that, do it as soon as possible, and be prepared to pay for it.

Part Two

Building the House

5

Out of the Ground

YOUR PROJECT PROCEEDS by getting permission to do so. No permits will be issued until you have a source of potable water. If there's no town water supply where you'll build, it's time to drill a well. A well, in the modern sense, is a six-inch-diameter pipe set into a hole in the ground. The pipe is driven into a hole drilled down to bedrock, or to a water-bearing layer, whichever comes first. Wells cost fifteen to twenty dollars per foot where I live, half that once the drill hits rock and stops requiring the pipe, called the casing. Picking the spot for a well is a gamble, pure and simple; it is usually located somewhere the huge well truck can get to easily. Wells have to be a minimum distance from septic systems, so a plan for the site is a must.

Once the well is done, the water often must be tested for coliform bacteria and other contaminants. Your contractor can arrange this. If the water checks out, you can leave the well alone until the plumber starts roughing-in the house. Then he'll slide an electric submersible pump down the well, dangling it from its delivery pipe. The pipe runs up to a fitting in the side of the casing, the pitless adaptor. This turns the water flow horizontal and connects to a pipe to your house, all below frost level. The pipe and wire come in through your foundation wall, to be hooked up to a pressure tank and pump controls in the cellar. Most water wells in my area cost two to three thousand dollars, with the pump.

With a plot plan and water test in hand, the builder can start applying for permits. How many and what kind depend on your state and local ordinances. The minimum is usually a building permit and a curb cut authorization. (A curb cut is where your driveway departs the road, whether there's an actual curb to cut or not.) If you will use the town's water and sewer system, you'll need permission to hook up to each. If not, you'll probably need a permit for a septic system, plus your water test. If your site is near a wetland or an ecologically remarkable area, you may have to get approval of your project from a local commission or a state agency governing environmental quality. Over a tight neighborhood may rest regulations covering the encroachment of a new building on an existing one, with certificates of compliance required. Most localities stipulate separate permits and inspections of systems for fuel, electricity, and plumbing.

All of which is to point out that this process takes some time. If your plan is rejected by the building inspector for any reason, it will have to be changed before you can proceed. The inspector must be sure all pertinent information is on your application. That done, he doesn't push too hard, as a rule, to see whether your plan is in strict compliance with the building code. I've never had a permit denied once the plans were complete. Theoretically, though, your builder could frame the whole house before an irregularity was discovered by the inspector.

From your point of view, it might seem that the inspector should check the plan for problems first. (A full set of plans and specs must be submitted with the permit application, and remains on file in his office.) On his side, the inspector knows most projects change between drawing and completion, so it's safer and easier just to check the building in sticks and shingles. It's up to the architect and builder to be sure the plan conforms to the codes.

Your contractor is in charge of rounding up all the permits, though you may have to sign some of them and make a phone call or two. Don't take out the permits yourself, because whoever does so takes responsibility for the work performed under them.

They sometimes cost plenty, too, and their price should have been included in the house bid. Municipalities look for revenue everywhere, and some building permits for single-family houses cost over a thousand dollars. Once all the permits are in place, the physical work of building can begin.

Excavation

Excavation — preparing the site for the foundation — comes first. The object is to make a hole in the ground of the right shape and depth, save certain material, and get rid of other material. ("Material" means dirt, right now at least.) Obviously, the kind of foundation you need determines the kind of hole you need for it.

Most residential excavating is done with a bulldozer and a backhoe, two quite different machines. Kids playing in sandboxes use a hand to dig a hole; this is the backhoe. When they use two hands to move a pile from one place to another, that is the loader. And when with hand and forearm they scrape a large swath, they are being bulldozers. So, details are picked out with the backhoe, and broad sweeps come from the bulldozer.

A typical small excavating business owns at least one bulldozer, one backhoe, and one dump truck, each costing around $60,000. These machines have useful, productive lives of perhaps ten years. The excavator also has trailors and pickup trucks, which last half as long and cost in the $15,000 range. And he has tools and garages, plus maintenance, fuel, and business expenses. All his costs must be accounted for in the hourly rates he charges for his machinery. So rates of $40 to $80 per hour are common.

This adds up to urgency at the job site. The excavator must charge for his time whether the bulldozer is idling during a discussion or pushing dirt. If we, the builders, have thoroughly prepared for the excavator's arrival with locating stakes, benchmarks, and clear drawings, we don't have to talk much. (A benchmark is a stake or sign in an out-of-the-way spot that denotes the exact height of the top of the foundation.) Some sites reveal surprises underground. The noise, shake, and smell of the machinery, as well as the high hourly rates, keep us on edge;

for handling the unexpected, preparation and experience are the therapy of choice.

For you, it's shock time. The lovely site you've walked so often becomes a theater of war. Pastoral visions flee before the dust and noise and pounding. Trees, the living tenants of your lot, are ripped from the ground. You'll be amazed how big a wound is made before the cellar hole is big enough. Any lingering anxiety you may feel over the siting of the house comes straight to the surface. Furthermore, below ground may lurk ledge (bedrock) or a spring to compromise your ideal basement.

The hardest part may be that this marks the real beginning of the job's passing from your hands to someone else's. In the planning stages, your desires rule. You've said, "I want that!" and it has appeared, better than life, in the plans. Now you must step back, or at least well to the side, to let those you've hired take charge. The brutish bulldozer undeniably shapes your new status.

Another unsettling development is the entrance of new characters. So far, you've dealt with a few builders and hired one you like, and the same with architects. You've visited real estate agents and banks. Everybody has more or less been in your camp, and your relationships with them direct. Here, now, is an otherworldly gang of heavy machinery enthusiasts. They aren't working for you but for some other guy who's working for your builder. They aren't beholden to you. They're polite, but you haven't much to say to them. They are wrecking your lot — in a good cause, of course, but with your money. There's just nothing you can do about this except be aware it's coming.

If your site has trees, the first thing the bulldozer does is stumping. Resist the temptation, even on a big lot, to bury the stumps off in a corner somewhere. This extends the scarring of the landscape and usually makes later work both necessary and difficult. Trucking stumps away is preferable. On big lots, we've hired a crew to come in with large machinery to snip off every marked tree and run the whole thing through a monster chipper. They even truck away the chips and sell them for fuel or landscaping. The lot is cleared almost instantly, and very neatly.

The next operation is stripping and stockpiling topsoil. The ground around the house will be abused during construction.

Any topsoil left there will be ruined by truck tires, excess concrete, and spills of lousy coffee. Because topsoil is valuable, a bulldozer operator scrapes off that fertile layer of dirt and piles it out of the way, to be spread around after the rough grading is completed, usually after the house has been sheetrocked. This is the ubiquitous pile of dirt you see at all construction sites. ("Grade" means ground level. "Rough grade" is the level of the dirt before topsoil is respread, "finish grade" the level after.)

Now the cellar hole is dug by the bulldozer. It is not a subtle machine, and digs a hole much bigger than the foundation size. The extra room is necessary for the foundation crew, but it also means a larger rent in the earth to be repaired in the finish grading. The most important thing we watch for is the depth of the hole

The footings (see "Foundation," below) must rest on undisturbed earth so they don't settle. Luckily, bulldozers dig only inches deeper with each pass. Sections of the foundation, for a garage or an ell, are often dug with backhoes, and we watch these more carefully for depth. A backhoe can quickly dig too deeply in a small area. The point of all this is to base the house on solid ground at the elevation determined by the benchmark. Inches count.

Some municipalities require a test hole during the planning stages, another good idea. Trouble sometimes appears when cellar holes are dug. The chance of a surprise can be reduced by digging a full-depth hole in the proposed cellar location. If you bought a lot already perc'ed, you can use the results of that test, or order another just for the foundation. This doesn't preclude problems in other parts of the cellar, but shows soil types and the presence of water or boulders below grade. Digging the test hole in the planning phase will help you get more accurate prices on excavation, foundation, and drainage work.

One form trouble takes is ground water, which simply means water found in the top ten feet of ground. Water usually travels in layers of sandy or gravelly soil between layers of less porous material, like clay. When you dig a hole and cut through a water-bearing layer, it's just like digging a hole at the beach: if you dig deep enough, water seeps in. Often as not, the water-bearing gravel holds only a small amount of water, which can be handled by proper footing drains (see "Keeping the basement dry," page 73). A larger flow should be stopped by a curtain drain ten or twenty feet from the cellar. A curtain drain is a stone-filled trench, slightly deeper than the water layer, that intercepts the bulk of the water before it gets to the foundation area. Curtain drains don't work if they can't slope to somewhere on your property; you may need technical advice on this score. Seek advice, also, on how much is too much water at the foundation hole. In all these matters, be sure to talk to your builder before consulting any outside experts. Ask his advice, get his recommendations, and involve him if you get outside help.

Another serious problem in a cellar hole is boulders or ledge. Unfortunately, these are missed in the test hole more often than is water. Ledge can't be moved except by blasting, which is costly and can wreck nearby water wells. We try to compromise with ledge rather than subdue it by force. We will move the house a little, make one end of the basement shallower, or pick away a corner of the ledge to fit the foundation in.

Boulders big enough to cause trouble for excavating machinery are rare. Huge ones may need to be moved by correspondingly huge machines. Just as bad is a layer of boulders several feet thick around the level of the proposed footings. These will make a clean, flat-bottomed excavation impossible. Often they have to be dug out and replaced with processed gravel. It must be installed in thin layers (eight to ten inches thick), compacted with a power tamper, and soaked with water before concrete can be poured over it.

You need to know all this stuff about dirt and rocks and water because your excavation should be done under an allowance. Your builder should state in his bid that the house will cost x, and of that amount, y is for excavating. If the excavating costs more, you'll be billed for it. If it costs less, you'll be credited with the difference. The allowance, like a partial stock-and-time contract, means the builder needn't inflate the bid to cover site contingencies, and that you must be willing to pay if trouble shows up. This is fair to both parties, and a reasonable way to handle problems beyond anyone's control.

An aside is in order here. When you start on the actual construction, in the dirt-work phase, you have most of your money and the least experience and caution spending it. This is one time when the natural flow of the house-building business works to your advantage. Everyone gets more keenly interested in the price of things when the money runs low. For this reason, most money difficulties happen near the end of jobs. I see this as a virtue, because the work done underground and under the house is the hardest to fix if done improperly. A leaky roof, skimpy insulation, even a sagging floor can be repaired fairly easily. A broken foundation, a wet cellar, or settling porch piers are way more trouble. So be glad you're spending good money for stuff you'll never see or think about again; that's the object. Your fancy fireplace mantel won't please you if it is tilting into the cellar on an unstable footing.

Foundations

A foundation is the means of transferring the building's loads to the earth. "Loads" means not just the weight of building and occupants and snow on the roof, but includes lateral and lifting forces from wind and differential loads from settling or earthquakes. A primary consideration in the structure of houses is getting the loads down to the foundation.

Foundations come in many types, depending on the loads they support. A stoop or small porch or gazebo, lightweight and open, needs only modest underpinnings. A porch that is attached to a presumably stable house must only be held up, and then just on one side. The house supports the other side and holds it against the wind. A gazebo foundation must anchor the building down and sideways as well as hold it up.

The design of many small and utility buildings is based on structural posts, and usually includes column foundations, called piers, below them. Piers may be concrete or pressure-treated wood, and should rest in the ground on a solid footing below frost depth. A solid footing can be poured concrete or a large, flat rock, depending on the weight of the building and soil conditions.

Concrete piers poured up to ground level are best, as they are strong and never rot. Pressure-treated wood rated for ground contact is OK unless the structure above is quite expensive or is an integral part of the house. In that case, you'll want the security of concrete. Assembled masonry, such as concrete blocks made into small columns, is weak and should not be used.

A step up from piers is called for when walls need to be supported more than at every porch post. This is the frost wall (more of a trench in southern areas), which goes down below frost depth and up to the bottom of the house. There's no cellar, so there's less digging and concrete—it's cheaper. Frost walls support the whole perimeter of the building, which allows normal (read: less expensive) framing techniques. The trouble with frost walls is what to use for the first floor.

The concrete slab floor is used mainly in warmer climates. Concrete can save you money, but it makes an unforgiving walking surface and is tricky to cover with certain other flooring. Slabs are sometimes tiled and used as thermal mass for solar-gain storage (see the "Solar heat" section of chapter 7).

What makes a slab cheap is that you can dig a trench the size of your four walls, fill it with concrete and pour more in the middle, and your foundation and floor are done. What makes a slab a problem is that it's tough to change. You have to run all your pipes first, before you pour, and there, by God, they stay. Even an insulated slab is very difficult to warm up from above, so cold feet are the rule—as are tired feet and legs from the unyielding surface. Linoleum helps, carpet helps more, but the sullen, hard thing sits there unrepentant.

The next type of foundation-floor is the crawl space, where a standard wood floor is installed over a knee- or waist-high cellar. A crawl space has the decency merely to be what it says it is, really only frost walls with the center hollowed out, and worthless for almost anything. The crawl space is now a mongrel but has a long history. For the settlers, a raised wood floor was a large step up from packed dirt. Today, with plumbing, wiring, and heating systems added, the crawl space is a poor choice. A plumber's most vitriolic stories are sure to include one of these damp, dark spider's nests, with lots of head bumping and cursing.

If you don't run into ledge or water, digging down the extra four feet to enclose a full cellar makes some of the cheapest space in a house. In many soils and locations, a relatively warm, dry, and light area the size of your first floor is yours for an extra four to eight thousand dollars. If you're forced by budget to build a smaller house than you want, this is money well spent. You'll get more space for less money than if you trade the cellar for a crawl space. The mechanical contractors will reward you with a cheaper, more reliable installation. And while a crawl space is useless, a full cellar offers many possibilities.

So what do you make the cellar walls out of? Poured concrete is quite common, especially in colder regions, and rightly so. Assembled masonry, such as concrete blocks, brick, and stone, is fine for holding things up, but not so good for enclosing a clean, dry cellar. Pressure-treated wood foundations are new enough so their life span can't be predicted. I wouldn't advise anyone to build an expensive house on an untested foundation. The other materials or combinations of them will work properly with enough preparation and attention to detail.

The great value of poured concrete for foundations is just the opposite: it will survive with minimum attention to detail. In most applications it is far stronger than necessary. Poured walls can support many, many times the weight they're normally loaded with. They are much stronger than assembled masonry against dirt pushing in from outside, the commonest cause of a failed foundation. Commercial forms and boom trucks make possible a straightforward foundation in three or four days. With a good foreman, a willing crew can push foundation jobs out one after another. With very little care, these walls can be made to last virtually forever.

A poured foundation, or any other, starts with a poured concrete footing. After making sure the excavation is level, the footing crew pounds in stakes around the perimeter of the proposed building. The stakes go in two rows, about eighteen inches apart, and boards are nailed to them, level all the way around. These boards will form the concrete poured between them into a continuous strip about eight to ten inches deep and eighteen inches wide. The concrete cures overnight, enough to remove the form boards, and your footing is in place.

The footing is always placed "below frost line." This means it sits deeper in the ground than the deepest possible penetration of freezing temperatures. Since moisture, and moist earth, expands as it freezes, a shallower footing would be subject to movement by frost. No building can take annual shifting with the seasons. Freezing under footings can move the biggest stone barn as surely as a fence post. In western Massachusetts, where Apple Corps builds, footings should be at least four feet below grade.

Sometimes we are faced with an excavation that has a weak spot. This might be a vein of water, a protruding boulder, or a hole left by an overzealous excavator. Since the purpose of a foundation is to provide equal support for all areas of the house, we like to compensate in the footing for unequal bearing. Usually we ask for thicker footings; sometimes we specify re-bar.

Re-bar, steel reinforcing rod, is used to hold concrete together. Concrete is very strong in compression, when it is holding some-

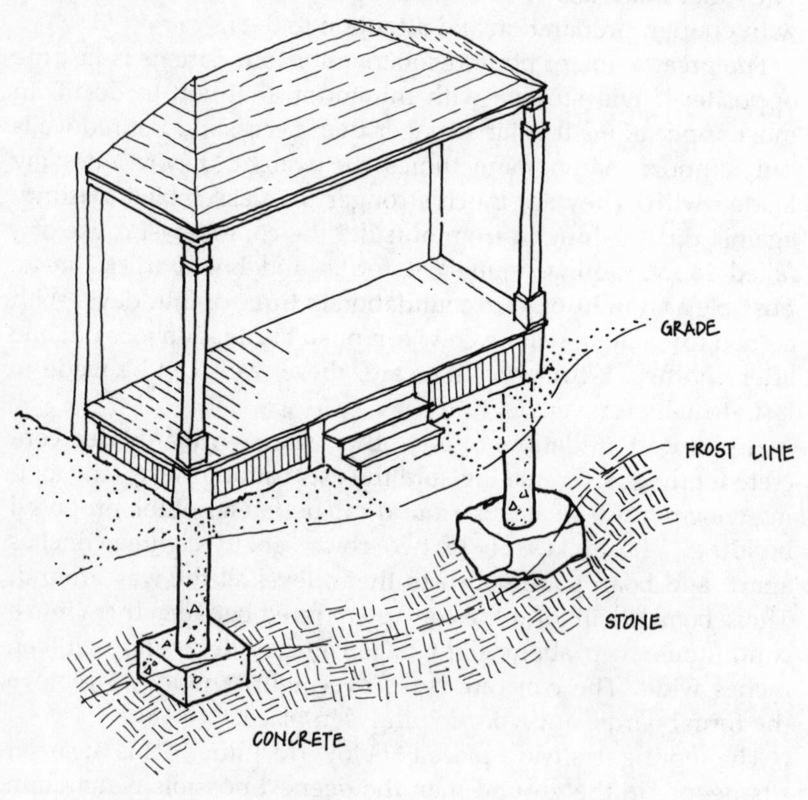

PIERS ON FOOTINGS

thing up. But if you had a concrete two-by-four, it would shatter if you dropped it or jumped on it. Concrete's weakness in tension, when being pulled apart or bent, can be counteracted by putting re-bar in it. Steel is strong in tension, and the concrete grabs on to the nubbly re-bar and holds itself together. So when we find that the footing must bridge over a much weaker (or much stronger) area, we put re-bar in it. This binds the under-supported footing to the stable areas on either side. Note that stronger is bad, too; the goal is even support.

Re-bar is also used in foundation walls. On a mostly level lot with good soil conditions and prudent foundation design, re-bar is not required. We like to see it when ground pressure on the walls is unusually great. Long, uninterrupted foundation walls are weak; corners make the walls more rugged. A lot that slopes down toward the house pushes hard on the uphill wall. A wet site means more pressure on walls, especially in freezing weather. Dense soils like clay push harder than lighter soils. Vehicle traffic near the wall creates a lot of pressure, and more traffic means more pressure.

A builder should, after a half-dozen years of experience, know enough to design a strong foundation. If you or he has doubts, an engineer can help. Trouble is, engineers often overdesign like crazy. They hate the thought of lawsuits, and who can blame them? Most architects' foundation drawings show re-bar. It is often included as boilerplate in the specs, but may not be necessary on your site. Reinforcing costs money you can spend on other things, but this is probably an area in which to tread lightly and go along with the plans. The general advice for all underground work is: overbuild.

A common foundation design flaw in cold regions is in the area of walk-out cellar doors or sliders. Such doorways are often cleared of snow, and repeated walking there compacts the soil, making it denser and heavier immediately outside the door. If the footings and foundation are not poured deeper under and near the door, frost can push this part of the wall every year. I figure on eight feet of twelve-foot-high wall for each walk-out door.

In busy years, the builder's main problem with foundations is getting one put in on time. Since everyone else has to wait until

STEPPED FOOTING FOR CELLAR DOOR

the foundation crew has done its work, the foundation contractor is very busy. Reusable forms are frightfully expensive, and the work is hard, so few builders pour their own foundations. In desperation, some resort to using block foundations, a poor idea. If your builder has a good relationship with a foundation man, you have a better chance of a timely pour. But your main weapon in this battle is preparation and advance notice. Getting your foundation in early is usually the most important step in getting moved in on time.

A good idea for a head start is to pour your foundation in the fall before a spring building start. Ideally, you would pour the foundation, wait three weeks or so for the concrete to cure hard, build the first floor framing and plywood it, and then backfill, at least partially. Spreading hay deeply over the footing inside and covering it with polyethylene will keep frost away. I don't advise pouring the cellar floor at this stage, because it would be difficult to protect from the weather. Since concrete cures slowly in the cold, you still must plan your fall pour ahead, but at least you'll avoid the spring rush.

Keeping the basement dry

Now you have a poured concrete foundation. The forms have been removed, and it's time to consider waterproofing and insulating the walls. Concrete is porous, even though it is dense. Water on one side will, before too long, get through to the other. Your excavation has created a natural low spot for water to collect in. Runoff from your roof or gutter downspouts concentrates water right at the vulnerable foundation wall.

Over 40 percent of new house cellars and 60 percent of existing ones leak water. Because wet basements are such a pain, there are many products available to combat them. I've always used asphalt coatings and have had no leaking problems *as long as the backfill was done properly* (I'll explain later). I apply roof patching cement in the foundation form tie holes, then brush on two coats of asphalt. After some years, this may eat away the styrene-foam wall insulation I also use; the jury isn't in yet. Another good system is draping thick polyethylene sheeting over the foundation. It works fine, but I've found it hard to manage amid the ups and downs of a cellar hole. Follow your architect's and builder's advice on wetproofing, because they're the ones who must stand behind it.

Foundation insulation is code-mandated in some areas, and it's a good idea in any case. As long as your basement is dry, it may as well be warm, too. I prefer exterior insulation to studding up walls inside and using fiberglass between the studs. I think of basements as utility areas and like the concrete showing on the inside. I also question the lasting qualities of wood right up against the concrete. I use one- or two-inch tongue-and-grooved styrene foam (extruded, not expanded) right over the wetproofing, and I tack it on the wall with concrete nails, one or two per sheet. The backfill holds it in place eventually. The foam shows above the finish grade, and must be protected by stucco or fiberboard, both with questionable life expectancies. Still, the exterior foam is mostly out of harm's way, and not too expensive, about six hundred dollars per square inch of thickness, depending on the house size.

To keep the basement dry, you must cope with ground water,

surface water (runoff), and roof and gutter water. Ground water occurs naturally, and was there before the cellar hole. If you dug your test holes in the spring, the time of high water tables, you should have found signs of it. The more ground water you have, the more trouble you'll have managing it. Most times, a small amount won't stop you from building. If a problem flow exists, in most cases it can be kept from the house excavation by a curtain drain.

Surface water comes from rain, melting snow, and drainage from other areas onto your site, and usually constitutes the largest amount of water you will need to control. Invariably, it is best handled at the surface, but first there are some rules. Don't build your house on a low point or pocket in the surrounding countryside. Don't build in the path of natural drainage from higher areas nearby. Don't plan anything—buildings, driveway, landscaping, plantings—that would divert natural drainage toward your house.

Do consider controlling surface water when planning your excavating. A swale, which is a broad ditch, will intercept lots of water, and can be incorporated into most site plans. If your driveway cuts across a wet area or seasonal water flow, you must install culverts to handle the water and a deep, porous base for the driveway. Most towns require that you provide a culvert where your driveway starts at the road, to cope with ditch water along the roadside. If yours doesn't, you probably should put one in anyway. It will likely save your driveway, not just the road, in a big storm.

Try to place your house on a high point, especially if your lot is wet. Consider raising the house higher above ground than usual, with the land sloping away all around it. (On all these matters a pragmatic builder will be your best adviser. If you can find a site planner or landscape architect with a streak of common sense, he can help, too. The fellow who is ultimately responsible for the job, your builder, is the one who should oversee all these decisions.)

The third source of water is the roof and the gutters. There are two ways to handle this water: disperse it on the surface away from the building or pipe it away underground. The dispersal system has a lot going for it. It is simple; you can see how and if it works. Most problems that may occur are noticeable and correctable. And it's usually cheaper than piping. The drawback is that even when it works, the water is still right near the foundation, ready to cause trouble. Plantings, settling, or landscaping work can alter the tilt of splash blocks, the common diverters used at the bottom of downspouts. This problem is often easily fixed, but someone has to notice it first.

The underground system is a network of buried pipes that the downspouts empty into. The pipes carry the water away to drain far from the foundation, a definite plus. And you needn't worry about the water washing out your lawn near the downspouts.

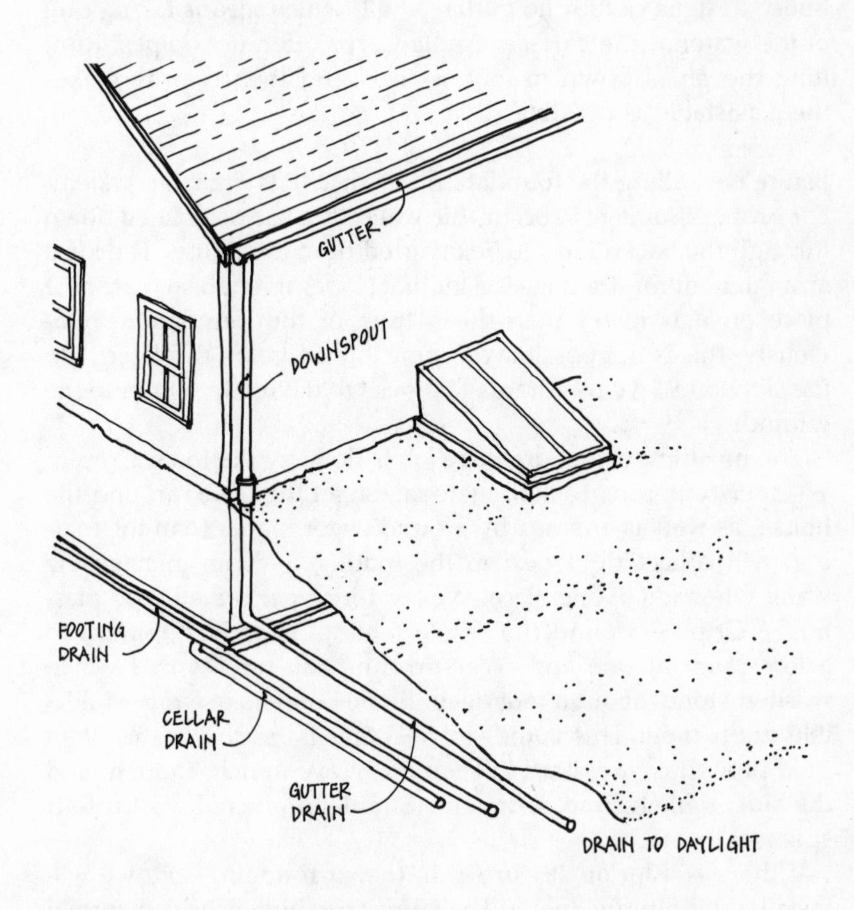

GUTTER AND DRAIN LINES

lems can come if the pipes shift during settling or freeze-w cycles. If the pipe joints come apart, water will leak out cisely where you least want it, next to the foundation walls. And if too much debris washes down from the gutters, the pipes may become plugged and water will back up in the same problem area. So, what to do?

Like many house-building decisions, this one is a judgment call. If the site is wooded or has large deciduous trees nearby, leaves in the gutters may make too much trouble for the pipe system. If the ground around the house is unusually well drained, surface dispersal will probably work fine. And some house designs call for no gutters at all, which means taking care of the water at the surface. Apple Corps often uses pipes. Running the pipes down to footing level and thence away makes them as stable as possible.

Before backfilling the foundation, we install its drainage system. For proper drainage to occur, the water that has percolated down through the backfill has to be diverted from the house. The ideal arrangement for drainage is a lot that slopes enough so that some place on it is lower than the bottom of the foundation. (Obviously, this is not possible on some lots; more on this later.) On the sloped lot, we run drainage pipes "to daylight," out onto the ground.

The most important drainage job is done by the footing drain. We specify porous backfill material, so all the water around the house, as well as any nearby ground water higher than the footing, will collect right next to the footing. We can manage the water when it's in one place. We lay 4-inch perforated rigid plastic (PVC) pipe around the whole footing, slightly sloping it to a low point at one end. We surround the pipe with 1½-inch washed stone, about a foot deep. Some builders use ridged flexible drain pipe. This comes in rolls and is much cheaper than rigid pipe. It is very hard to keep on an even pitch, though, and the ridges likely trap sediment. Stipulate 4-inch PVC in your specs.

If there is a lot of clay or silt in the surrounding soil, we will install a soil-filtering fabric. The fabric goes under the stone, then up over the stone and pipe, and partway up the foundation wall.

The goal is keeping the drain pipe from becoming plugged with sediment that has washed down through the backfill. If the surrounding soil is coarse or gravelly, we just shovel a few inches of pea stone (3⁄8-inch diameter) over the 1½-inch water-carrying stone. The pea stone keeps the backfill material from infiltrating the larger stone, but lets water through.

Now we have a continuous pipe surrounding the footing. This pipe is connected with a T-shaped fitting to an unperforated pipe. The new pipe heads for the nearest daylight that allows it to slope at around one quarter of an inch per foot. The farther away you must go, the lower must be your destination. The end of the pipe should be covered with galvanized wire mesh to keep critters out, and protected by a small stone arch to prevent it from being crushed.

On some lots there simply is no such place. In that case, we run the footing drains inside through the footing, to the cellar and into a sump. (A sump is a built-in low spot in which water gathers.) We set a chimney tile — a two-foot-long, foot-square terra-cotta pipe — into the cellar floor, surrounded below by stone. All the water ends up in the sump, and an electric pump, started by the rising water level, pumps it out and away from the foundation through a pipe. Any mechanical system is subject to problems, so it's better to use the drain-to-daylight technique wherever possible.

It's not time to backfill that drain pipe trench yet, though. Another pipe, running from a drain in the cellar floor, has the same destination. We usually place one floor drain near the bulkhead, in case someone leaves it open on a rainy day. If the laundry or boiler or water heater or pressure tank is nearby, so much the better. For servicing and catastrophes, we want to be sure any inside source of water is near a drain. So if the water heater is thirty feet from the bulkhead, we'll put in two drains.

The floor drain should empty into a pipe separate from the footing drain's, running all the way to daylight. If they are joined, a cheaper way out, a clog downstream of the joint will cause the footings to drain onto the cellar floor! Similarly, the gutter drains, when used, should run down a separate pipe. So we often have three pipes running down the ditch. The bottom of the ditch must be fairly even, so puddles in the pipes don't t sediment and block the water flow. Finally, we cover the pipes by hand with the first eight inches of dirt, to make that rocks tipped into the ditch by the machinery don't p cture the brittle plastic.

Use the right backfill

Back to the hole in the ground, with foundation. To make sure surface water gets down to the footing drain, we pay close attention to backfilling the foundation.

The quality we seek in backfill material is porousness, which allows water to pass through it rather than soak in. That means space between the particles, a feature of light soils. Soils in which most of the particles are similar in size don't make good backfill. These include clays as well as fine sand. The trouble with fine, even-particle-size soils is that the bits can pack tightly together. Surface tension allows moisture to stick in the matrix, and the result is a wet, unstable mess, a guarantee of trouble to come.

The alternative is gravel, which means different things to different people. To me it is a coarse, light material with many different sizes of particles. It appears as a mixture of sand and rocks, from grain to bowling-ball size. Variety is the key. We get gravel from natural deposits or from a rock-crushing operation, where it is processed and graded for certain characteristics. In that case, the range of particle sizes is smaller, but the water-shunning quality is similar.

In coarse, mixed-particle material, larger grain sizes and uneven spaces between grains allow water to drain through. Air breaks up the surface tension. You don't have to be a soils engineer to determine what material you have on your site. Soak some with water, then squeeze it in your hand. If it stays in a clump when you let go, don't use it for backfill. Even a little. Ever.

Ideally, we could use the dirt we dug out of the cellar hole for backfill. We wouldn't have to truck it away or find an on-site dumping place, so you'd save money. More often than not, though, we must truck in gravel and truck out the diggings. This is what makes backfill a pain in the neck. You have all this dirt right there; why not just dump it back in? Because it's just too

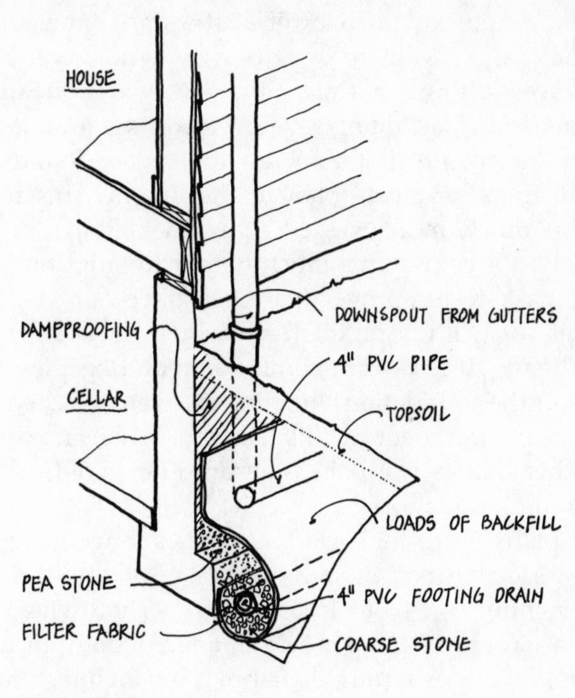

BACKFILL

expensive to dig it back out again a couple of years later trying to get the basement dry, or the porch to stop heaving, or the walk to stop flooding.

The coarse backfill material should start right over the drainage stone. It should extend three or four feet from the foundation and come to within a few inches of the finish grade, leaving room for the topsoil. If the excavation is large, the loader operator may dump a bucket of gravel against the wall, then a bucket of diggings outside that, to save gravel.

If you plan to have brick or stone walks or dry-laid steps, there should be two to three feet of gravel under them, too. A garage floor should have a minimum of a foot of gravel under it, compacted as it is placed. (Compacting means running over the material with a gas-powered vibrating tamper each time a few inches of gravel are added.) Watering the material as it is dumped in helps to settle everything. We are trying to re-create undisturbed earth, and it takes some doing.

also important to backfill at the right time. Just as we foundation walls to be strong, we know we can reduce forces acting on the walls. The vibration and added weight caused by backfilling severely loads any foundation wall. But if we install the first floor on top of the foundation, we essentially brace one side of it to the others. This makes the foundation much more rugged; now backfilling can be done more safely. Of course, it's much harder to build the first floor when we have to jump over that deep ditch outside the walls. Sometimes we will fill the ditch partway. This is a compromise between being able to work comfortably and keeping the main pressure off the foundation. Juggling excavators' schedules with our own — having them backfill halfway, leave, and then return for the other half — is a pain. We can often avoid the problem with a strong foundation design.

Taking pains with the backfill ensures a long-lasting foundation. It's perfectly easy, though, to do a quick-and-dirty job and have everything come out looking good. The owner, thinking nothing is wrong, is happy. The builder, money in hand and hoping nothing goes wrong, is happy. The architect most likely wasn't there during backfilling and doesn't know anything was amiss. Two or three years later, or maybe not until a particularly severe winter, the wall may crack, piers lift and tilt, and the brick steps slide away from the door.

Now the foundation is dampproofed and insulated, and the drainage is installed. We have filled around the walls up to about a foot from the top of the foundation. We have placed gravel in the areas under walks, steps, and driveways to a depth of two feet or so. We have used some of the dirt we dug out of the cellar hole, but we still have a lot left.

It's tempting to spread this dirt around the foundation, tapering it away to leave the house on a little mound. While this is good for drainage, it can wreck the looks of the place. Great slabs of foundation exposed above ground shows poor planning or a builder saving excavating dollars. The more the house can appear to have been placed artfully *in* the existing landscape the better. The final height of the building is established by the benchmark, set at surveying time. If that means there's dirt left over, so be it. It's usually easy to find someone who needs the fill.

Before we dug the cellar hole, we "roughed-in" the driveway. Now we smooth it out again for the next months of tradesmen traffic. We shape the land around the building with a bulldozer, to conform with the site plan. We leave the level of the site slightly below where it will end up, to allow for the thickness of topsoil. We overfill any ditches or deep excavations, knowing they will settle substantially over the next weeks. We make sure we leave no stumps, branches, or trash buried around the site.

Before going on, let's imagine your first disagreement with your contractor. You have come over to watch the rough grading and notice the bulldozer operator burying some stumps. You think he shouldn't do that, but what next? If your contractor is there, go talk to him. Since burying stumps was specifically banned by the contract, you'd like him to talk to the machine operator and have him stop doing it. He should comply, but suppose the contractor isn't there? Or worse, suppose nothing was said in the contract about the stumps?

You might be reluctant to approach the bulldozer operator yourself, especially if the contract says nothing about stumps. He may seem an alien creature, far removed from anyone you're used to talking to. Even when you get his attention and he stops digging, he has the upper hand. He knows what he's doing and you aren't sure. To underscore his position, he leaves his machine running (this is normal for a diesel) so you must shout — use his style, if not his language. He's working for the builder, not for you, so can you tell him what to do?

Usually these issues can be settled, or at least put off until the builder can intervene. The operator may be able to see your position, somewhat. In most cases, he won't want to make trouble for himself or the builder who hired him by continuing a contested course of action. But the best defenses against the awkwardness of this situation are a complete contract and a builder who is on the site. Everyone will know what to do and what to expect.

If you did all the right things ahead of time and the builder tells the bulldozer operator to go ahead and bury the stumps over your objections, then you chose poorly. Fire your builder right now. Your contract should have a section that tells you how to do that. Much will speak against starting over, but it is inevitable that problems will get more serious and harder to fix as the job progresses. To avoid another confrontation, you may keep quiet about things that bother you. Gradually you'll become worn down and the project will turn sour. You'll get a worse job that will take longer and cost more than you thought.

At this point the septic system, or sewer connection, goes in. Then the well or town water line is connected to the house. We make sure any buried services are in place. These might include phone lines (with extras for home businesses or kids' lines), cable television, intercoms between buildings, yard lights, and so on. The goal is to leave the house site generally clean and smooth, but not finished. The immediate area will be trashed as the house is worked on; there's no point in getting fancy now. We want to be able to work without obstructions, and it's time to start sawing boards.

6

Wood at Last

THE FIRST LAYER of wood in a house is called a sill, or mud sill. It establishes the size and shape of the building, and holds it down as well as up. It's bolted to the foundation, and the house is spiked to it. Since foundations are often slightly off level, the sill gets shimmed up at each anchor bolt to bring it all to one level. The shims are wood shingles, and bristle alarmingly around your new foundation. The exposed bits get cut off later.

We always use pressure-treated lumber for sills. Sills are among the first things to rot in a house, and should be protected. The conditions for rot are a combination of dampness and oxygen; all that's needed is a light covering of autumn leaves. In old country houses that were constructed low to the ground, leaves and debris build up the height of the soil around the foundation, often above the sills. We've replaced enough sills to know: always use treated wood. The next layer of wood, the floor framing, is considered far enough above the causes of rot to be safe. In termite-prone areas, a metal shield is often placed between sill and concrete, but it must be big, and kept clear of pipes and wires, to be effective.

Inevitably, between foundation and sill there is a gap, perfect for winter winds to penetrate and chill the house. To fill this space, we usually install a fiberglass sill sealer before bolting down the sills. All sill sealers are a pain in the neck. Mainly, they interfere with the shims, and make the height adjustments more vague. They are not thick enough to fill a large gap of one inch or more. They are not attached to anything, so they may slide out or be pushed out by a shim. And they seem likely to absorb and hold water, bad news if the pressure treating lasts just so long. We have tried getting around them by injecting foam between concrete and sill after the house was framed up, but didn't think we could get enough in to do any good, especially in narrow gaps. We still use sill sealers; we just grumble a lot.

Now that the perimeter of the house is ready to build on, we can consider the middle of the building. In older houses this is where sags show up in floors; central supports were often too weak or too few. Floors must easily support anything you put on them, either the family spinet or forty dancing partygoers (several tons in motion!). House floors are made from regularly spaced joists with plywood on top. Houses or sections of houses between twenty and about thirty-two feet wide usually have one beam, called a girder, under the joists. Girders are placed often enough to keep the unsupported joist length (span) under about fifteen feet. With larger joists this span can be increased to eighteen to twenty feet. (You certainly don't have to learn about joist spans to hire a house built. But you will want to know what you're looking at while your house is going up.)

Girders should be the massively strong members their name means to most people. Girders are installed from end to end of the building or section they support. There are two keys to their strength: span and depth. The span is the length of the girder between supports, and the depth is the distance from the top to the bottom. Rules of thumb aren't very useful here, but expect to see girders made up of four two-by-tens or two-by-twelves set on pipe columns eight or so feet apart.

The pipes, known as lally columns, are filled with concrete and are very strong as long as they sit on a big concrete footing, three feet square and ten inches thick. They can't sit just on the concrete cellar floor, which won't take the concentrated load. They sometimes make for inconveniently broken-up cellar spaces, but the girder must be supported. The lally columns can

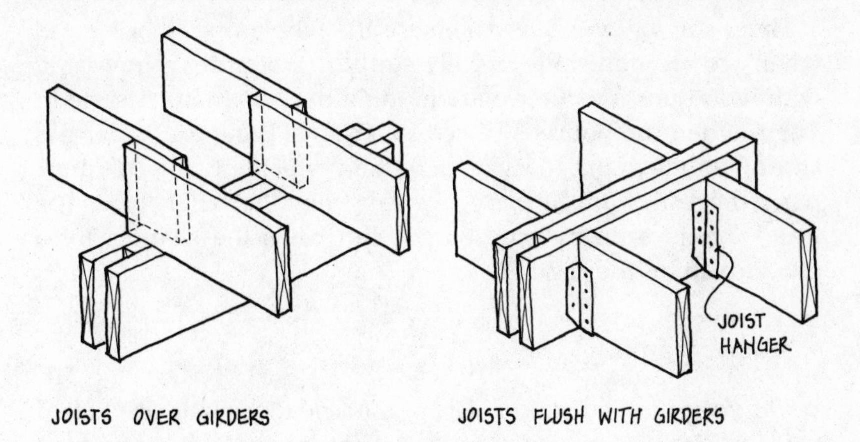

JOISTS OVER GIRDERS   JOISTS FLUSH WITH GIRDERS

GIRDERS

be eliminated or reduced in number by replacing the girder with a wall, if the bottom is made of pressure-treated wood.

Girders can be placed flush with (even with the tops of) or below the joists. Flush girders make for more headroom in the basement, but mean attaching the joists to them, usually with metal joist hangers. Putting the girders below means the joists rest on top, making them easier to install. The lower girder also means less basement headroom.

All framing lumber shrinks after it is installed, especially after a heating season. Normally, the joists shrink more than the girder. If the joists are flush, shrinking sometimes makes a broad lump in the floor above. This is particularly noticeable in a timber-framed house, where the girder is one massive piece and the joists are smaller and lighter. The girder twists and shrinks a little, the joists shrink more, and whoop goes the floor. If the girder is below the joists, it can be shimmed up to bring the floor level and leave no lump.

There's another problem with a flush girder. In order to get necessary pipes and wires into a central wall, subs must drill into the girder from the sides and then down from the top and hope to meet somewhere inside. Or subs will hack a great chunk out of the side of the girder to get their pipes lined up under the wall. Either way, it's a difficult job, one that weakens the girder and costs you more.

There's a way we can make sure the whole first floor is level when we are done. We usually support the girder temporarily with wood posts while we are framing the rest of the first floor. Then, when the house is closed in, and we think the girder has shrunk all it's going to, we install the permanent lally columns. After the lallies are in place, we can pour the cellar floor. The wait is good anyway; rain or hot sun ruins the chances for a good finish on the concrete.

Floor framing

So far your new house consists of a foundation with sills bolted to it and one girder, sometimes more, running down the middle. The next step is the joists, which form a grid that supports the whole house. Sills and girders are below, the house above. The house rests on the joists where the walls are. The joists are also the support for your floor. Over the joists goes plywood; builders call the whole thing a deck. This is all part of framing, a term you'll hear a lot and should understand.

Framing describes all the structural parts of the house, like girders, joists, studs, and rafters. It also includes the skin that covers all these parts, made of plywood or fiberboard or foam, which is called sheathing. It does not include any finish materials, like clapboards or drywall or cabinets. Framing is also the generic term for making a structure for anything; as such, it may apply to cabinets and bookcases. In house carpentry, though, it is the noun standing for the structure, or the verb describing the work being done.

A framing job on a house starts with the sills and girders, and proceeds through floor and wall and roof structures and sheathing. Also included in a framing job, though, is putting up the roof trim and shingling the roof, and often installing doors and windows, and maybe even siding. So the framing *job* includes all these operations, but the framing *materials* include only the structural members and the sheathing. Back to the deck.

Joists are most often two-by-tens, sixteen inches on center, but what does that all mean? "Sixteen inches on center" means simply that the joists are installed sixteen inches apart, measuring from the center of one to the center of the next. This spacing is

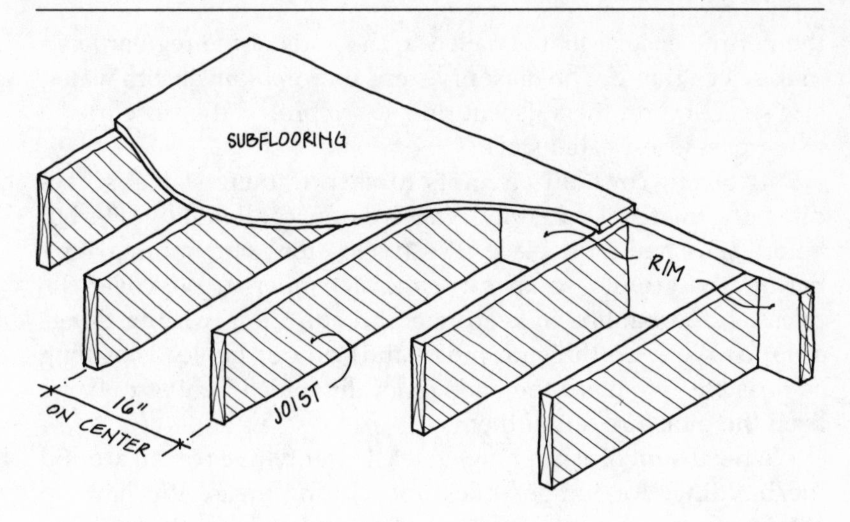

JOISTS

standard because it allows the easy use of many modular building products, such as plywood and insulation. Three sixteen-inch spacings equal four feet, and six equal eight feet. "Two-by-ten" refers to the thickness, then width, of a joist (or any other framing member). It is usually stated along with the stick's length and written "2x10-14" (two-by-ten, fourteen feet long). Two dozen of them would be written "24/2x10-14."

When woods of different species are combined in a deck, the contrasts in their capacity to absorb and release moisture affect the stability of the deck. If placed side by side, joists of two species may shrink and expand unevenly, and leave you with an uneven floor. The ideal to aim for is a floor (or wall or roof) structure built from all the same species, all the same moisture content. So, rather than go to a stronger species for a long joist span, I'll specify a bigger size of the same wood.

The exception to the rule of sticking to the same species is when we need pressure-treated framing for a certain area. Most pressure-treated lumber is southern yellow pine, an unruly wood I avoid for framing. I've already discussed the pressure-treated sill used to avoid rot. I also install pressure-treated rim joists where masonry, such as brick front steps or a poured concrete porch, will fit against the wood deck framing. A rim joist is the perimeter joist that is nailed to the ends of the regular joists in the deck frame. The masonry steps or porch can absorb water, and soak and rot the adjacent rim, so we protect the structure by using pressure-treated stock.

Part of our constant attempts to keep framing stable is handling the material carefully after it is on the building site but before it is nailed in place. We try to store our framing lumber off the ground so air can circulate under it. We cover the piles with tarpaulins to keep rain and sun from twisting things around. We keep the strapping bands on our big loads as long as possible. We paint the ends of locally milled lumber to try to keep the moisture out of them.

We need a lot of room now, at least twenty-five feet all around the building, for lumber piles and cutting areas. We have to allow five or six feet all around next to the house for the wall and roof scaffolding. We know we'll get new deliveries of lumber before we finish with the old piles, so we plan on double- or triple-size areas for storing materials. It's great if the lumber trucks can get all the way around the building, so new cutting areas can be opened up while the old ones are still in use. It takes something like a twenty-by-twenty-foot area for each cutting operation, even bigger for joists and rafters.

Now, let's get this deck built. First off, we check the sills to be sure they are faithful to the original dimensions of the building as specified in the plans. Meantime, someone else will be laying out. This term refers to marking, on sills and girders, the places where each of the framing members is to be installed. These locations are all critical, since they define the openings for stairs and chimneys and laundry chutes, as well as provide support and nailing for the deck plywood. For consistency, we choose one corner of the house and make all our measurements from there. The layout marks will show the joist installers where each piece goes. And they tell the cutter the length of the headers that support the various openings. (Headers "head off" the joists from running into various openings. Headers generally are structural members running perpendicular to the adjacent framing, defining the edge of an opening like a stairway or window. In a floor, a header can be on either side of an opening, and in a wall forms the top of the opening.)

Before we nail any joist down, we sight down its length to see whether it is crowned. A crown is a bend that would make the joist belly the floor up or down. What we're after is having all the bends, or crowns, upward when the joists are installed. Any weight on the floor will tend to straighten the joists out; if they were installed crown down, weight would only bend them further. Areas where framing accuracy is most crucial, like the rims and around staircases, get the best and straightest joists. Any joist with too big a crown, say 1⁄300 of the length or more, we reject, cut into shorter pieces, and use as headers. This means that a sixteen-foot joist with a three-quarter-inch crown is at the limit of acceptability.

Now we've nailed all the joists and headers to the sills and girders. Each joist laps, in a straight-fingered handshake, its opposite number, assuming the girder lies below them. We nail the joists securely together at the lap, and nail what's called blocking, or solid bridging, made of pieces of joists, between the ends to keep the joists from twisting. We occasionally knock these blocks out later to accommodate plumbing or wiring, but that's easy. We can usually move them if we know ahead of time where the pipes will run.

Next we nail in the rim joists. These are our straightest pieces,

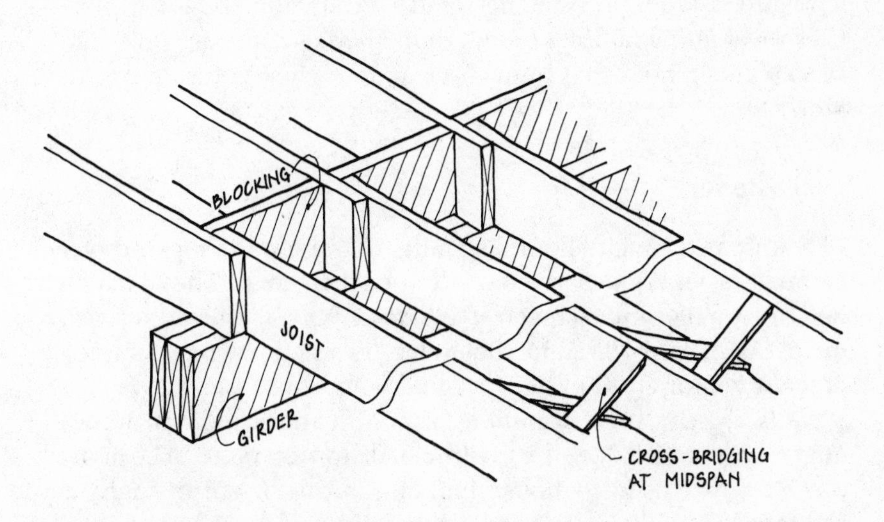

GIRDER BLOCKS

for they form the outer edge of the deck, the perimeter of the house proper. They also serve to hold the joists upright and to forge the separate joists into a unit, a solid deck equal to its heavy responsibilities. We often increase the solidity of the structure by installing cross-bridging — short, diagonal struts nailed between the joists from the top edge of one to the bottom edge of the next.

Cross-bridging is stronger than its alternative, solid bridging, because it can be wedged into place, making a tight connection between the joists. Each pair of bridging pieces joins its joist to the next; a straight line of bridging across the center of the joist span stiffens the floor considerably. Bridging also helps spread the weight of a heavy object like a piano or refrigerator over several joists for more support. Cross-bridging struts must be nailed in at their top end before the deck is sheathed. The bottom end of the bridging should be nailed only after the sheathing is in place, and preferably just as the house framing is completed, to allow the joists to settle into their permanent positions. A modern, and thoroughly inferior, alternative to wood cross-bridging is metal bridging. The metal struts sometimes vibrate and hum, shriek if they rub each other, and, worst of all, don't wedge the joists solidly like their wood counterparts. And they look lousy. We who bear the standard of bygone building practices revel in the failure of modern methods to improve on them. Who wants a floor that hums back at you when you stamp your foot?

Fasteners

I haven't said much about fastening all these parts together yet because fasteners deserve a reading of their own. They include not only nails and screws of all descriptions, but adhesives, metal brackets, bolts, and clips. Here is a nail primer, a survey of what you might find lying around your house site.

Nails are the most common fasteners, and a modern house might be stitched together with a half ton or more. The most common nail used in house building is called, surprisingly, a common nail. It's made from a short piece of plain wire with a point on one end and the other end bashed flat to make a head.

This nail is used for most structural connections, and h
virtue of ubiquitousness; everyone knows what to expe
a nail. Nails come in different wire diameters, each with
dard corresponding length, so that a tenpenny nail always is about one-eighth inch in diameter and three inches long. Easy to understand, cheap to buy, easy to use, and it works; truly a miracle product.

Like anything simple, nails have been gussied up over the years to suit different purposes, sometimes well, sometimes not. Nails have been coated with various substances, twisted or stamped to look and act like screws, and adorned with different heads. They have entered the high-tech age, queued up like machine gun bullets on a plastic cartridge belt, ready to blast from a nail gun, a sort of automatic hammer. Common nails have stood the test of time, but modern construction techniques offer other choices and many improvements.

Coated nails are the next most useful in a building. One standard coating is galvanizing, named for the way the zinc coating used to be applied to the steel nail. The only kind of galvanized nail worth buying is one called hot-dipped galvanized. The alternative, electric galvanized, has a smooth zinc coating that is not well attached to the nail wire, and skins off if the nail bends or hits a knot. Hot-dipped nails have a resilient, thick zinc jacket that withstands more of the perils of a nail's life. Galvanizing keeps nails from rusting, and the crusty coating on the hot-dipped number makes it the very devil to remove. For this reason, we use hot-dipped common nails to fasten sheathing to framing all over the house, floors, walls, and roof. We also use them for exterior trim and decks and stairs, and for installing some kinds of windows.

Hot-dipped commons have a skinny little brother, kind of a pest, called the hot-dipped galvanized box nail. The common galv nail is just what it says, a galvanized common nail. The box nail is a common nail made from thinner wire so as not to split the thin slats of wooden boxes. They don't split wood as readily, but they bend so often and so abruptly that no one wants to use them, except for clapboards and thin box slats.

Hot-dips also come as finish nails, those nearly headless varieties used on trim so the heads don't show as much. We've found that for exterior trim, a finish nail, galvanized or not, just won't hold any substantial board on tightly if the weather has other things in mind for it. The unsightly big head is the very thing that grips the board, so we use the brutes and explain our reasoning to those who ask. It's better to have to explain the heads than why the board is falling off the house.

Another surface used on nails is cement coating. This is a thin film of glue that sets when heated by the friction of the nail forcing its way through the wood fibers. Cement-coated, or c-c, nails are usually box-nail size and so bendy that they're thrown as often as driven. Besides, a galvanized coating grips better than the glue. A cousin of the c-c box nail is the sinker, a thin, sharp framing nail with a little glue on it. Sinkers are pretty easy to drive and stick better than common nails, and they cost more, of course.

Threaded or screw-nails are designed to hold better than common nails. They may be twist-threaded, like a wood screw, or annular-threaded or ring-tail nails, depending on whether the threads spiral along or form parallel grooves in the nail wire. Pole barns, often made from unseasoned lumber, are assembled with big threaded nails to hold the building together even as the boards dry and shrink. Stainless steel ring-tails are the nail of choice for clapboards destined for a clear finish. And tiny underlayment nails, ring-tail box nails, will hold down a plywood subfloor even if the joists below twist a little. Threaded nails are terrible to pull out, and often break off, like a lizard's tail, rather than be recaptured.

A simple variant is the duplex, or form, nail, a common nail with an extra head above the regular one. The duplex is used to assemble rough, temporary structures like scaffolding. The second head allows you to drive the nail home and still have something sticking out to grab on to when it's time to pull things apart.

In some applications, screws are superior to nails. Screws can fasten two boards together without anyone beating on them. A screw will make a strong connection even if there's nothing solid about either board — a job beyond the reach of a nail. An example of this is fastening two thin strips of wood for the facings of a bookcase to each other before attaching them to the bookcase. Any frustrated amateur nailer knows how hard it is to nail two flimsy things together. In fact, some of what looks easy about a pro's nailing is aided by his propensity to nail only into something solid.

I also use screws when even a little loosening of the attaching point will be hard to live with. Sheets of drywall, big and heavy yet tender, I attach with screws; any loose fastener in a sheetrock ceiling always shows and is bothersome to fix. Subfloors (the first layer of sheathing over floor joists) under tile and linoleum get screwed down, to form the most stable base possible. And I always use screws to install cabinets and built-ins. Screws make a very solid connection, and they can be removed and reinstalled many times. No one screws framing together because it's unnecessary for strength, takes longer, and costs more for screws and labor.

The development and availability of hardened, thin-shank screws has simplified many carpentry tasks. These screws are easy to drive mechanically, as they have an accurately formed Phillips or square drive head that keeps the driver from popping out. Their thinness makes it possible, in many cases, to drive screws without drilling a hole first, so busy carpenters are more apt to use them. Originally developed for drywall fastening, and often called drywall screws, they are indispensable for that task. They are even available galvanized now, though the coating is so thin they rust outdoors. Along with construction adhesives, they are expanding the possibilities of good building practice.

Construction adhesives, usually applied with a caulking gun, are getting more use than ever. Subfloors are glued down and then nailed, paneling is glued to walls, and foam insulation is glued to plywood and to concrete. One reason for the widespread use of adhesives in framing comes from the mechanical uncle of the nail and hammer, the nail gun.

The nail gun is powered by an air compressor. The nails come in rolls, feed into the machine one after another and are driven into the wood by a piston. The piston is light, in order that it move fast, and pushes only on the nail. This makes sense mechanically but differs from a hammer. When you drive a nail with a hammer, the last blow hits the nail and the wood simultaneously. This forces the nail into the wood and forces the wood

nst whatever it is you're nailing it to, all at once. So when
ı nail something with a hammer, you fasten it firmly. The nail
n drives the nail at high speed but with little mass, so the nail passes through the wood without moving or even disturbing it much. Inertia keeps the wood in the same place it started out. If that happens to be tight against the next board, good, and if not, then it stays loose.

This is not a big problem in some framing situations, but definitely is when attaching deck sheathing. If the plywood is not fastened tightly, then when you walk on it you will force it down against the joist, and when you move away it may spring back to where it started. Plywood moving down and up the shank of a less than fully driven nail makes the dread floor squeak. In an old house, the squeak comes from a slightly loose nail that over the years has let a floorboard rub up and down its shank to squeak with every motion.

So the nail gun helps carpenters go fast, but causes noisy floors. Enter subfloor adhesive, squirted on each joist just before the plywood is laid. This sticky stuff grips the plywood and the joist and joins them fast. Some of the time gained using the nail gun is lost, but the floor is undeniably solid. A subfloor nailed down with our favorite hot-dips benefits from gluing too, since every so often a nail will loosen up as the joist it pierces shrinks. The life span of construction adhesives is uncertain because they are a relatively new product, but they surely seem to demonstrate the benefits of advances in construction technology.

We use adhesives anywhere we won't tolerate noise or motion. We fasten stair treads with screws and adhesive, though a friend uses silicone caulk for the same purpose, on the theory that it retains more flexibility over the years of wood movement. We stick on foam insulation panels with foam adhesive; nails don't hold crumbly foam very tight. And we use two or three quarts of plain old carpenter's glue in every house, spread through most of the finish woodwork joints.

Metal brackets show up in our houses now and again. We prefer, probably from stubbornness, wood-to-wood joints, but metal connectors work better sometimes. The most common connector is the joist hanger, a bent-tin stirrup used to carry the end of a joist. Hangers are also made double-wide, and we often use these to support a header in a tight spot in the floor framing. Hangers, though galvanized, are apt to rust in damp settings, but then so are nails. Floors over crawl spaces or in other damp areas should have their joists notched over ledgers, beefy wood strips spiked to the girder with galvanized nails.

Heavy timber-frame designs sometimes call for correspondingly large metal brackets to join the timbers. Some post-and-beam houses built from kits depend entirely on this system, anathema to the wood craftsman. Angle iron and bolts can join two beams, but can't make a pleasing structure. They try to substitute steel for subtlety, crudeness for craft, bolts for beauty.

Small, H-shaped plywood clips support adjacent edges of plywood sheets between roof rafters. They allow slightly thinner plywood to span the same distances without buckling. They work, but the plywood ought to be thick enough to make a solid roof anyway, so why bother? Steel can substitute for plywood as bracing for the walls of a house, as diagonal straps nailed to the studs. Leaving the plywood off causes other problems, though, which I'll discuss in the section on sheathing in chapter 7.

Generally, the advances in construction technique and fastener development have been directed at cutting costs, not producing superior results. I believe in the long-term economies of a solid structure. I trust what I know works, not the plastic present, for guiding my way through the building of a house.

Decking

When all the joists and rims and headers are in place for the first-floor deck, we sheath it. Deck sheathing is more commonly called decking (both noun and verb) or subflooring. The subfloor has four jobs. The most obvious one is providing a floor for the house, or, more accurately, a support for the finish floor. (A finish floor can be carpeting, tile, wide pine boards, and so on; it just means the topmost layer of floor, the one you walk on.) The sturdy subfloor provides a flat surface for the carpenters to work from. It also ties all the joists together, making the deck a structural unit. Finally, a subfloor braces the deck frame diagonally, cooperating with the foundation in providing a rigid and permanent base for the house to rest on.

Back in the previous chapter, I mentioned decking the first floor as a method of bracing the foundation walls against the pressure of the dirt outside. While the deck framing braces the foundation somewhat, sheathing the deck in plywood makes it solid, supporting the whole plane of the deck at once. The sills are bolted to the top of the foundation, the joists and rims are nailed everywhere to the sills, and the plywood conjoins the assemblage into a stiff and strong unit. Plywood is the key to the deck.

Plywood is a product of technology that I wholeheartedly, even automatically, accept. Construction plywood is made from peeled logs cut, as you might meticulously peel a cylindrical apple, into broad, thin sheets called plies. The sheets are stacked up so their grain runs perpendicular directions in alternate layers, with glue between the layers. The sandwich is pressed tightly until the glue hardens, and is then trimmed to size. The number and thickness of the layers determine the thickness of the finished sheet, usually from a quarter of an inch to an inch or so.

Peeling logs into veneers (the plies of plywood) has been possible for a century or more. The recent technology lies in the glue between the plies. This stuff is strong, easy to use, waterproof or moisture resistant where need be, and cheap enough so manufactured boards can compete with the natural variety. The glue isn't perfect but plywood is worth a few problems. The main trouble is that the glue sometimes doesn't hold the plies together uniformly, so they delaminate (come apart) in sections, and the plywood becomes weak and uneven. The cheapest plywood is made with the poorest glue, so we don't buy plywood by price.

Plywood has two great virtues. It comes in big pieces, so each covers a lot of floor (or wall or roof). And the way it is formed makes it very strong in the plane of its two large dimensions. Because the plies cross each other at right angles, each reinforces the natural strengths of its neighbor. Plywood, securely fastened, fiercely resists forces pushing and pulling on its edges. It makes a wonderful brace for framing.

Decking with thicker than average plywood makes a better subfloor. Its inherent stiffness tends to equalize the differences in the crown of the floor joists. It acts a bit like bridging as it ties adjacent joists into a sturdier whole. And it glosses over minor variations in the installed height of joists, girders, and headers. resulting in a flat subfloor. So how thick is thick enough?

Three quarters of an inch is right for living rooms over joists sixteen inches on center, and for bedrooms over twenty-four-inch centers. One-half inch is the traditional standard and works OK; three-quarter inch is twice as strong. There is middle ground here, five-eighths inch, but it always seems the black sheep of the plywood family: poorly glued and sized, and full of defects. Of course, the thicker sizes cost more; three-quarter is half again more than one-half inch — say, eighteen dollars versus twelve. Someone will have to decide what to use, though, so you might as well be informed of the choices.

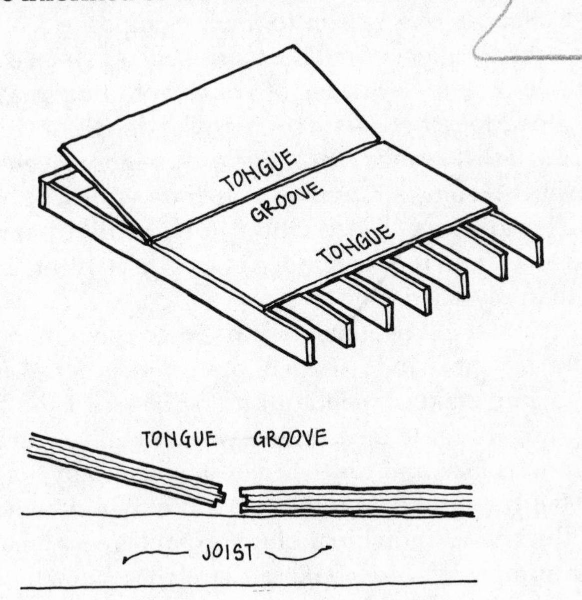

TONGUE-AND-GROOVE PLYWOOD

Three-quarter-inch plywood (virtually all construction plywood is four feet by eight feet, so plywood is described by its thickness) also comes tongue-and-grooved. A tongue, a slight protrusion, is milled along one edge of each sheet, and fits into a corresponding groove in the adjacent sheet. This ensures that the sheets will not shift, relative to each other, where they are not supported by a joist below. The ends of the sheets are cut straight, because ends of plywood sheets must land on joists.

I've had problems with t-and-g plywood, though. Any plywood exposed to moisture will swell a little, getting a bit wider and longer. For this reason, it is good practice to leave a little space (about one eighth of an inch) between sheets used on a subfloor or roof that might get rained or snowed on before it is covered. But with t-and-g plywood, you have to jam the sheets together to make the tongue fit in the groove. Tight-fitted plywood can stretch a carefully built deck frame a half inch in a light rain, more in a three-day soaker. The standard sheet is sturdy enough between joists, and, when spaced, avoids the swelling problem.

This swelling is one reason to avoid composite materials generically called flakeboard. These are sheets, shaped like plywood, made of chips or flakes of wood from fingernail to palm size and glued together — sort of a phyllo dough of wood waste products. In most brands, the chips are roughly layered, which gives the sheets some strength and their other descriptive name, oriented-strand board. Since OSB can be made from the small chips left over when real lumber is cut, it is 10 or 20 percent cheaper than plywood.

I never use it. First of all, since it has more glue than plywood — to hold all those odd-shaped pieces together — it weighs more. The glue makes it solid but not stiff, so it is both heavier and weaker. The glue also dulls saws and other cutting tools quickly, a nuisance and an expense. It's also ugly, made from scraps. Mainly, though, it swells up more than plywood, especially in thickness. Individual chips expand more than others, making a lumpy surface. OSB can stretch a house an inch or more, the sheets will buckle in the rain, and the nails will come loose when the sheet settles back down after drying out. As trees are cut down faster than we'd like, economical use of all the scraps makes environmental and business sense. But OSB doesn't cut it as sheathing.

When I started in this business, most exterior plywood was fir, a dense, stable, and strong western species. As fir veneer logs (you need a good log to make plywood from) got scarcer and more expensive, manufacturers used more substitutes, like southern yellow pine. This species is fast growing, and is often planted and harvested on tree farms. SYP is a temperamental and unstable species, though, and the plywood from it is heavy and obstinate, and buckles badly in the rain. Pine plywood is about 20 percent cheaper than fir, and is often the only kind available. Still, it's worthwhile to shop around for fir.

Finish floors of various thicknesses should end up even with each other so you don't trip as you walk around your house. Oak strip flooring in the dining room is about ¾ inch thick, and can be fastened right to a solid subfloor. The wall-to-wall carpet in the family room might be 1¼ inches thick, but goes right over a good subfloor, too, and compresses underfoot to the level of the oak. Vinyl flooring in the adjacent kitchen, about ⅛ inch thick, needs ⅝-inch underlayment to come up level. And the powder room tile is around ¼ inch thick, for a ½-inch underlayment.

Underlayment is a special sheet used as a base layer under finish flooring. Thin floorings such as vinyl need an underlayment that is smooth and solid, so high heels or skinny stool legs won't crush it, and stable so it won't buckle or swell up if it gets damp. Particleboard underlayments, finer versions of flakeboard, are smooth and solid, but should never be used in kitchens or baths, where moisture can push them around. Plywood comes in an underlayment grade called PTS, for plugged and touch-sanded. That means defects in the top two veneers are cut out, and the voids plugged and sanded level. If the top veneers aren't solid (the second veneer, or crossband, rarely is solid in regular construction plywood; the first sometimes so), then you can't put vinyl on top. PTS is stable, and often specified for under tile in dry locations. (Unfortunately, I've found that PTS often does have gaps in the second ply, making it a questionable product for use under sheet goods like vinyl. The manufacturers' associations should correct this mislabeling.)

Carpeting doesn't need a particularly great underlayment; the padding takes up minor variations, and the backing layer of the carpeting itself spreads out concentrated loads fairly well. Particleboard is OK for this, though there are health questions about the glue used in making it. There are foam plastic and fiber products for carpet underlayment that offer some sound deadening and low price, but I've never used them. I stick with PTS under sheet goods and a combination of subfloor and underlayment under carpet.

Lately, though, I've been using three-quarter t-and-g fir with a built-in expansion joint at the groove. This is marketed as a combination sheathing-underlayment, good under carpeting or wood and strong enough for any residential use. If your budget is tight, go with half inch, but the stiff floor is more satisfying.

7

Walls and Up

THE DECK IS securely fastened to sills bolted to the foundation. With plywood laid, crews can work on a flat surface, which is easy to sweep and mark up and build on. The backfilling is finished, and the area around the house is graded smooth, though not ready for the grass seed. Good planning means the septic system and the utility trenches have been installed, so there won't be new ditches to trip over while we're putting up the rest of the frame and completing the outside of the building.

We will pick up the site so when we start the walls, a new operation, there will be few remnants of the previous one. We'll rake up wood scraps, roots, and rocks so we don't trip while we're carrying heavy planks. Roots and rocks can often be buried in later excavating or trucked away for fill, and we pile wood scraps for use as kindling in the owner's fireplace or stove. If we have leftover lumber from the first-floor deck, and don't need anything like it later in the construction, we return it to the lumber yard or store it away from the building, under cover. Many lumber yards discourage returns by charging 10 or 15 percent of the goods' cost for handling, though the fee is usually waived for good customers. Experienced contractors just don't return much stuff; they know what to order, and they often find a use for surplus.

Placement of doors and windows

The real work of erecting the first-floor walls starts with reading the drawings. The architectural drawings we use, no matter who drew them, are called the plan view, or floor plan. This presents a bird's-eye view of the first floor, with the rooms labeled and the outside and inside walls shown, the doors and windows indicated but not described, and some interior features shown and/or described. Some architects include the size and exact location of windows and doors on the floor plan, which helps.

Locating and detailing the openings in the walls of a house is one of the major jobs of the designer. Though I have no formal design training, I have learned from many architects and customers, and built lots of buildings. I, not your architect, am the one who gets the Sunday phone call when rain tumbling from the gutterless roof and flooding the flush deck has soaked the floor inside the slider. Avoiding trouble makes me pragmatic, conservative, and skeptical. As referee, I favor Function over Form when they compete. My hope is that your architect will be able to back both contenders to the detriment of neither. This is a tall order, so I try to help whenever I can.

A careful architect considers much more than aesthetics when he designs a house's façade. Weather holds a house in a tenacious and ultimately fatal grip from the day construction begins. Traffic flow inside and outside the building has much to say about how the house can be used. Air and light must pass through the house, or its passing be restricted, at the choice of the tenants. Windows and doors, being the most expensive areas of the conventional house shell, and the most used, deserve a lot of thought.

Doors first. Doors must be easy to open and close, to get through, and to pay for. They must close tightly against the cold, and open with a light touch. They must appeal to our sense of being welcomed into the house, and resist attempts on them by the unwelcome. They should be easy to install and easy to maintain. All this points us directly to a hinged, insulated steel door.

Steel doors are available in many styles and in a limited number of sizes. The better ones have soft but tough weather stripping on the sides and top, and an adjustable threshold to make the bottom draft-free. The inside and outside steel layers should be separated by good foam insulation, with wood fillers around the edges. I like the Peachtree Avanti series — simple, cheap, and well built. When properly installed, this door will last many years, even under a regime of neglect. Even if you select a fancier wood door for a formal entrance, a steel door is a necessity for your high-traffic side door.

Outside doors must be protected from roof runoff, wind, and rain. I consider the minimum protection to be a one-foot overhang not more than two feet above the top of the door, with a gutter to divert the runoff. Almost anything extra you do to keep the weather away is money in the bank. An actual roof, breezeway, or recess over your door is best.

Whatever you stand on just outside the door must be lower, preferably a full step lower, than the door's bottom, to keep water outside. This rule applies especially to decks and steps under the unguttered eaves of a house. Rain bouncing off the deck or step saturates the door, frame, trim, and weather stripping. This constant soaking is tough on all parties, and if the door gets much sun as well, it will suffer all the more. In the snow belt, built-up snow and ice can actually funnel water right under the best weather stripping and into the house. And just try getting out of the house after a snowstorm when the storm door has to push a foot of snow ahead of it as it opens. A flush deck is a common design flaw on the second floor of houses and condos.

Many house designs highlight the entranceway, and well they should. In fact, any house design that hides the "front" door from a visitor's first glance should be thrown out immediately. You must reassure your guests that they can find their way to you without pause or doubt. Hiding your door in a cute nook or cranny may be intellectually interesting, but it's an affront to the visitor.

Certainly the doors should get you in and out of the building in the right places. A conventional front door placement ushers you into your living room, the side door into the kitchen. In

…se, a foyer, vestibule, or mud room makes the transition … tside to inside slower and more awkward, but more ther… onscientious. Wintry air is trapped in the small room, and … ching summer wind is baffled before it reaches your easy ch… Guests, too, may appreciate a chance to get used to their new surroundings for a moment before they enter the heart of your house. Don't make the path of the grocery carrier too long or crooked, though; you have to live here, too.

Most exterior doors swing into the house. This allows for a storm or screen door outside. You can't really use the floor space the door swings into, but the door itself may be in the way when it's standing open for the breeze. The open door welcomes you and your guests in, but makes you step around it to get out. An outswing door opening onto a screened porch is a great entry. You'll probably end up with an inswing door because of convention, but there are choices.

Resist if possible the urge to install a Dutch door, fashioned so that the top and bottom halves open independently. This unit has twice the hassles in installation and maintenance of a conventional door, and it isn't available in steel. It's rare that some other choice won't serve as well.

Some house designs call for a sliding glass door to be used for normal house traffic. This kind of door makes a terrible entryway. Sliders are hard to open and close, as the motion calls on weak muscles. They don't latch securely, as the standard mechanisms are awkward and flimsy. The track at the bottom gets clogged with debris easily and makes the door difficult to push. Constant use quickly deteriorates the weather stripping, and sliders let in lots of cold air. They're great for occasional access to a deck or patio, and admit plenty of light for the dollar, but they are poor doors.

A new version of the slider is the patio door. Like traditional French doors, the patio door is two door units in one double-wide frame. Unlike French doors, the operating door often is hinged between the pair, and the lockset is at the edge of the opening. The patio door is a good compromise between a slider and a single swinging door. It lets in a lot of light like a slider, but moves and latches and seals like a swinging door.

Lots of our customers, though, like the authenticity of French doors. Traditionally, both doors operate, though one can be easily latched and the other used as a single door. Wood French doors are often drafty, as the twists in each door tend to be magnified where the two come together. Storm doors are a must in cold climates, and are often worse fitting than the primary doors. Some folks, including me, love the look and feel of the traditional doors, and put them in against all advice. Romantics have their French doors, and pragmatists their snug dens.

Tradition holds less sway in the placement of windows in new houses. Windows used to march across buildings in regular order, spaced the same distance apart, lined up over doors or over each other. The Victorians fooled around with window placement, along with everything else, but their variations usually followed a pattern. In many contemporary designs, randomness seems a virtue, order criminal. Contemporary means anything goes, and windows are shot in scatter-gun fashion at the façade. The more expensive or elaborate the house, the more liberties the designer takes, a paradox of expected upper-class conservatism.

I think your house should either have a traditional window and door placement or clearly show reasons for its alternative layout. Rooms have to be ventilated, and windows are, so far, the most reliable method. You'll want light during the day and privacy at night, winter sun when it's there but no blank, black holes to chill your evening moods. You'll want your house to reflect your desires and your personality. A wild and brave new design may showcase your open-mindedness, but I don't need the mental exercise of rationalizing the myriad and conflicting details. I like symmetry, or at least order, to define the solid base from which I venture forth to conduct my life. More power to you, though, if you really want to live in a radical statement. I'll even build it for you.

In any case, in any design there are rules to ensure the endurance of the house shell. Don't place a window in the path of splashing water (say, from an adjacent ell roof). Don't stick a window in an end wall so close to the ell roof that you can't get a few inches of siding, with its water-turning flashing, in be-

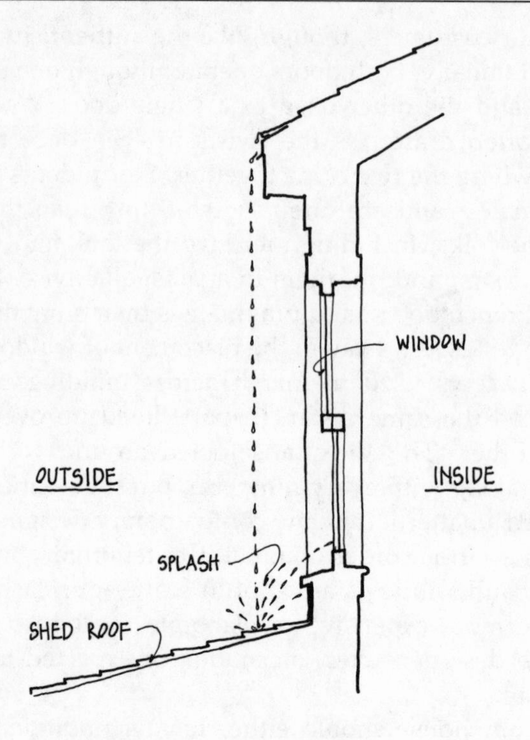

WINDOW AT SHED ROOF

tween. Similarly, if the roof of a porch or sunroom lands just under a second-story window, the juncture may cause trouble.

Never, never put a window in a tub or shower stall. Condensation on even triple-glazed windows will deteriorate the window quickly, and cold downdrafts will spoil your baths. Running the shower curtain all the way around the tub is a big drag, and ruining the window because you didn't is, too. A much better solution is to rearrange the room so the tub wall is opposite the window. You may have to change much of the adjacent floor plan to get there, but it's worth it. No matter what kind of window you put in the shower, it will be nothing but trouble.

Even a large and fancy spa or hot-tub room must have a place for condensed water from the windows to run without causing damage. A cold winter day on one side of the glass and a steamy tub on the other mean quarts of water running down. Tile the sills and the floors below, and slope everything to a drain. This is a serious amount of water, which can't be contained by wadding up towels on the window sills. It's not just bathroom windows that cause trouble with water damage. Any window, but especially a casement, left open on a rainy night can admit destructive water into the structure as well as the interior of the house. We try to put them under wide protective overhangs.

Solar heat

Overhangs bring up the burdensome though many-sided subject of solar heat. I have looked at many "solar houses," built some, and repaired a few. Most have glaring problems.

First, they have an unusual concentration of mental and financial resources focused on the solar design, generally to the detriment of the rest of the house. You have to live in the house summer and winter and day and night. Ipso facto, a solar design emphasizes only one aspect of what your house means to you and does for you. Within the confines of a strict budget, allocating money disproportionately means your house won't be a unified whole, balanced in what it offers you. You can't spend an extra $15,000 on what you may call a heating system without shirking on $15,000 worth of shingles, sinks, or staircases.

Second, many solar house owners have trouble with their houses, though not all will admit it. Solar gain has been such a powerful deity, since 1973 at least, that many misguided pilgrims have marched to the Temple. Houses were built as experiments in the "new" technology, and many owners have been baked and frozen, showered and snowed upon, albeit in the true pioneer spirit. The most frequent problem seems to be overheating of solar-heated space during warm weather. Common sense says that a big glass wall facing the sun will overheat the house on a hot summer day, and all the thermal mass in the world can't keep up with it. Designing the house with only winter in mind is like designing it as if it were always night.

The means and methods of successful solar design are not beyond the reach of a conscientious designer, but the design (and often the construction) of many solar houses has been by fervent idealists of narrow focus. A dramatic statement, rapturously embracing the dogma of innovation, is not my idea of a house.

A third problem, concomitant with the first, is that of the disregarded aspects in the design. Solar gain is achieved through south-facing glass, but what if the best view is to the north? What if the logical entrance is also to the south? Will it be obscured in the shimmer of glass? Are you to be so oriented to the south side of the house that the northern rooms are dark, stale, shadowless recesses? A successful design will integrate all the areas, neglecting no useful space.

Such is the strength of the proponents of solar design that they don't need me to stand as advocate for it, though I believe in many of its tenets. I preach instead for an overarching balance in house design. Shucking tradition to clear the mind, and slate, for new designs and technologies is an intellectual exercise and should be cautiously applied to the places we live in. A house is not a commodity to be purchased and discarded if it doesn't fit. It endures, if it does, as a chunk of your neighborhood, your country. You have broad rights and freedoms in your house design, and narrower but stronger duties of stewardship to your community and species.

In any case, a rundown of solar-gain problems should be helpful. The main one seems to be in controlling the gain and harnessing it for use. Merely putting lots of windows on the south side gets you a house too hot even on a winter day, and too cold on many nights. Insulating the windows at night is difficult without bulky and awkward devices. Shading them in the summer is required lest the occupants should become brownies in a solar oven. Interior shades are not too effective, and exterior shades must be selective, admitting winter sun and shutting out summer's. And exterior shades must be built to withstand the rigors of climate, the same as any other part of the house shell.

Thermal mass is touted as the solution to the wild fluctuations in gain and loss that make the house uncomfortable. Thermal mass means having heat-absorptive materials in sufficient quantity, and in the right location, to soak up heat during peak periods of solar gain and dissipate it at night or on cloudless days. Thermal-mass solutions are good to the extent they do their job. The materials must be able to store a lot of heat, must accept and release it in good time, and must be in accord with the décor of the house.

Brick or concrete walls, barrels or tubes of water or chemicals, and to some degree tiled concrete floors are all used for thermal mass. I think the mass should not fly in the face of the rest of the house design, nor should it block a pleasant view. Many thermal-mass walls indeed block direct sunlight from the only rooms that would get it, a real travesty. The exact amount and location and composition of the mass should be determined by a competent engineer, for these are not easy features to move or change! It's rare when floor storage mass works effectively; if your floor can't stay at, or at least average, your approximate skin temperature, you'll be cold in that room.

Do make sure to leave yourself some windows you can walk up to and look out, if there's something outside you want to see. In the country or the suburbs especially, being able to see the driveway or parking area from several places in the house (including your bedroom) is reassuring. It's nice to see outside even without a spectacular frame of distant peaks, so consider your best views when you orient your house and locate the windows. Don't be a slave to Solar South and disregard the other qualities of your life at home.

Wall framing

I'm itching to actually build walls now. The first task is to figure out how high they are and where the openings will be. The height is usually easy to figure. What we need, though, is another drawing, called a section, which is a stand-up look at the house as if it were cut with a cleaver. This view shows the innards of the construction, as well as details of the walls the cutting reveals. It also shows us how high the walls are, and how the walls and the floor and the roof framing are intended to fit together. Along with two or three additional drawings, the elevations, the sections tell us the height and width of the various window and door openings, and their exact locations in the walls of the house. Accuracy in the framing makes the rest of the house go together smoothly, so we take pains studying the drawings.

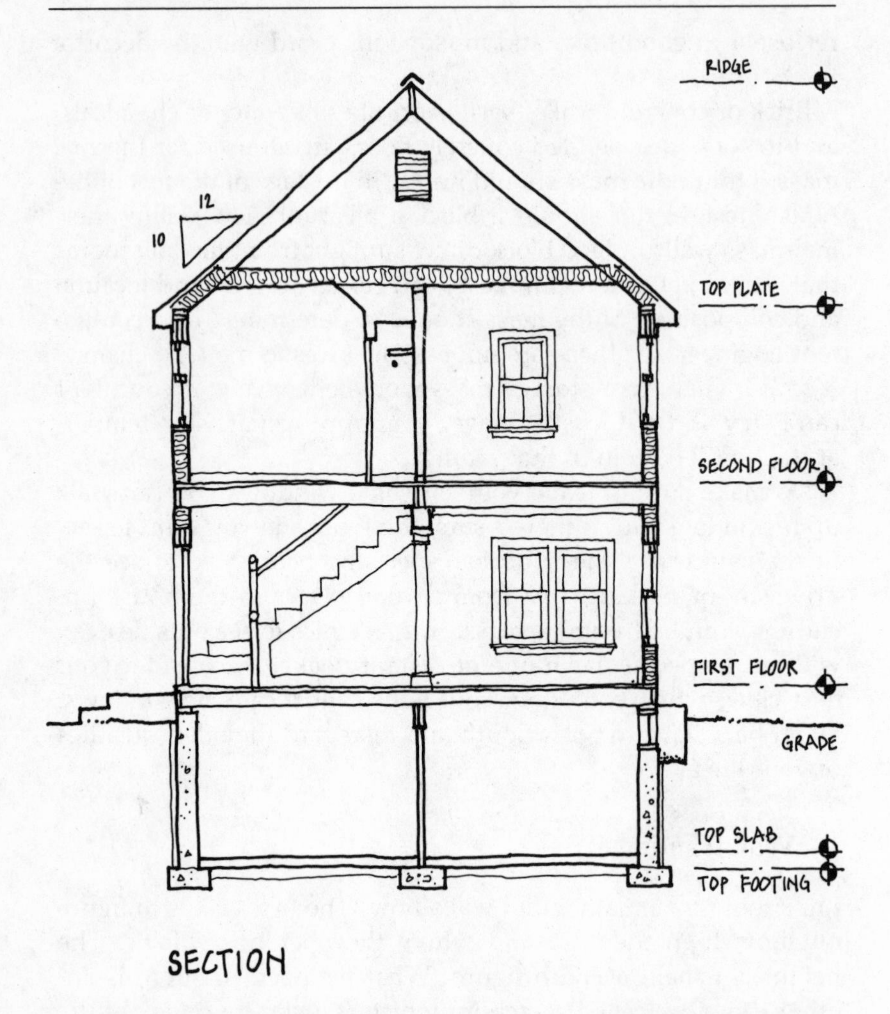

One thing to keep in mind when figuring the height of doors and windows is whether the tops line up or not. Often drawings show an elevation, a straight-on look at the outside of the house from each side, that has the windows and doors aligning. On a house with clapboard siding, this means the same run of clapboards rests atop the windows and doors. In order to achieve that blessed state, we need detailed sections of the doors and windows themselves.

Before going ahead, though, it's time to learn some more terms. First, a rough opening, something you'll hear a lot of as your house is framed. "Rough work" applies to the framing of a building, and that includes the openings in it for doors, windows, chimneys, and so on. So a rough opening is a hole in the framing. Rough openings need to be more precise than their name suggests, as they determine the exact final placement of whatever fits in them. Different manufacturers' doors and windows, and different styles from the same manufacturer, require different openings — the reason changing your mind even during the framing means redoing the rough openings.

A rough opening in a wall usually consists of a header supported by jacks. A header forms the top of the rough opening, and is the beam used to hold up the wall or roof above. It is

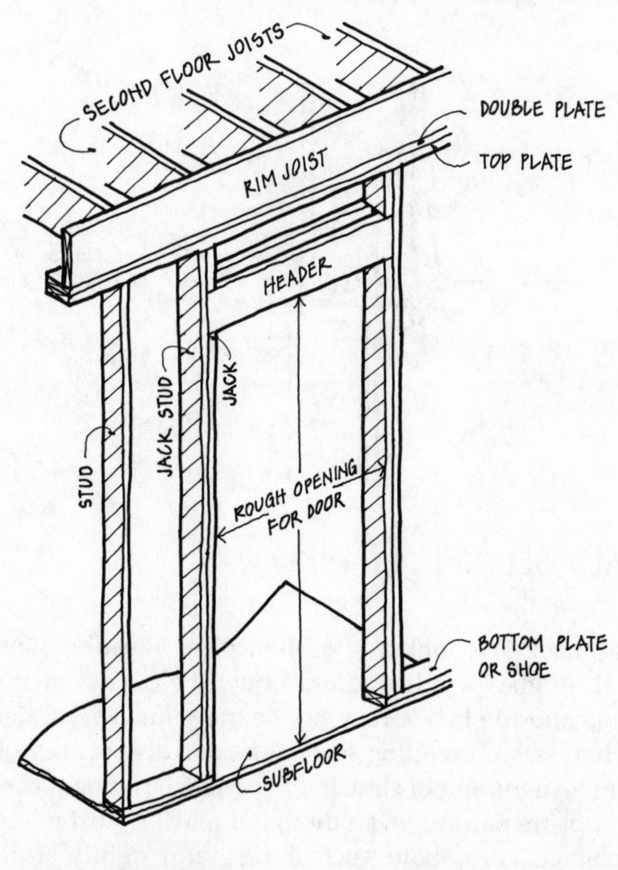

WALL FRAMING

often not one solid beam, but is built up from smaller pieces. At either end the header rests on jacks, studs that take the loads from the header down to the deck and thence to the foundation. Window openings also have rough sills, which hold up the window unit, and cripples, short studs that support the rough sill. So doors and windows have jacks holding up their headers; the windows rest on rough sills; the doors rest on the subfloor.

Studs are the orderly and regular vertical pieces that form the walls' structure. They provide places to nail on inside and outside surfaces, spaces for insulation and wiring and plumbing, and support for the floors and walls and roofs above. Studs are fastened at their top and bottom to plates, long horizontal pieces of the same material and size as the studs.

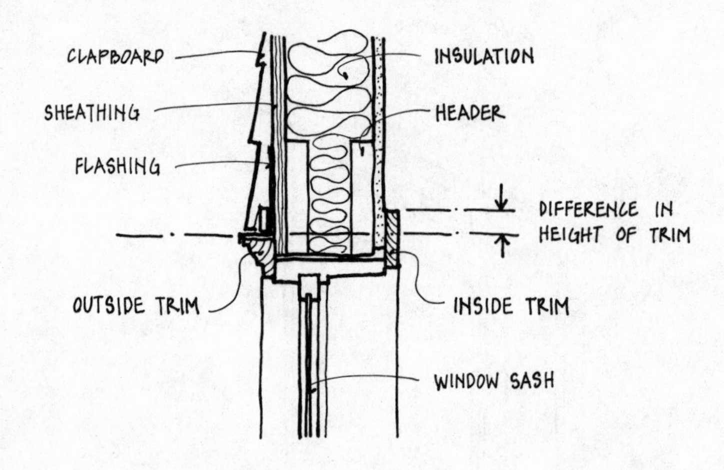

WALL SECTION AT WINDOW

Let's go back and look at a section view of an outside wall, running through a window. Notice how the casing, or trim, on the outside and the inside may not be the same height, depending on what is used on either side. The same discrepancy obtains with doors. Alignment of similar objects in a building pleases the eye and appears natural, even though it may be hard to achieve. Most architects care about such things, and rightly so. If it is impossible to make door and window casings align inside and out, they usually opt to match the exteriors, because clapboards will show the differences plainly and painted walls will not.

If you're wondering why your builder stares glassy-eyed at the drawings for long minutes when you think he should be sawing and hammering, this is part of the reason. Architects try to make sure all the elements of their design fit with one another, and your builder checks, too. If a window won't fit where it's drawn, he shouldn't build it that way, even though responsibility for the design lies elsewhere. It's part of his job to see that the project is buildable. Many architectural drawings stipulate that the builder field-check some measurements. "Verify in field" means the architect, in his office, can't be sure of the accuracy of a detail.

Next, we read the door rough openings on the drawings: 3′-2 x 6′-10½. Rough opening measurements always show the width first, then the height. Three foot two, as we say it, is thirty-eight inches, but we use the feet-and-inches convention because it is less cumbersome when measuring bigger dimensions. Most architects separate feet and inches with a hyphen, and omit the inch marks.

The details of the window head and door head (the top of the frame, or jamb) show us that the outside window trim is a half inch lower than the window rough opening, and that the outside door trim is one inch higher than its rough opening. So we'll build the window opening an inch and a half higher than the door opening, and everything will line up, outside at least. Variables in this simple calculation may include: raising the door rough opening to gain clearance for a thick door mat; dropping the window rough openings so other architectural features, like a beamed dining room ceiling, won't come down over the glass; and the whammy that comes from unusual door or window shapes, like the now popular "round heads."

Once the height alignment decision is made, we can plot the rough openings from left to right using the plan view. Most complete floor plans show the distance to the edge or the center of openings and intersecting walls from the corners of the building. As when we're plotting the deck layout, we always try to start from one designated corner of the house, so everyone working will know the point where the measurements began.

Thus, the center lines of the window rough openings might be 4′-6, 9′-0, 18′-0, and 22′-6, measured from the northwest corner. Sticking to this convention, like the one using feet and inches, reduces the chance of errors caused by random math mistakes.

Once we know the sizes and positions of the various openings, we're ready to begin laying out, marking on the top and bottom plates the location of each opening and each stud in a wall. Laying out requires thoughtfulness and precision, and is usually done by the foreman or an experienced carpenter. The marks he makes show the framing carpenters not only where to nail all the studs but their lengths, and the sizes and lengths of other wall parts — headers and jacks and cripples and sills.

One of the uses of wall framing is to provide nailing, something solid to nail into, for all the surface materials. So we make sure to install, in the rough framing, nailing for partitions (inside-wall framing) where they intersect the outside walls, for fancy or unusual inside or outside trim or moldings, for most cabinetwork and some shelving, for certain light fixtures, for dryer vent caps, hose faucets, and so on. We try to anticipate in the framing every place something will be attached later, and ensure there is solid support for nailing in each instance. Anyone who has a loose outside faucet on his house understands the logic for making sure there is something solid to attach his new one to.

Almost all the houses Apple Corps has built in the last ten years have two-by-six studs for the exterior walls. They make a thicker wall, with room for more insulation, than the two-by-fours that used to be the standard. Since two-by-sixes are sturdier than two-by-fours, we can place them twenty-four inches on center rather than the sixteen inches that two-by-fours require. Using fewer studs means that two-by-six wall-framing costs are similar to those of two-by-four walls. Since each stud creates a break in the insulation blanket, fewer studs are better. (More on this in the insulation section of chapter 12.) We don't, however, trade sturdiness or thoroughness in framing for the chance to leave out a stud here and there. The corner of an outside wall is a good example.

The corner framing of house walls must be sturdy. Outside, the corner boards (the trim running up the corner) need solid nailing to stay put. The ends of adjacent clapboards should be nailed to the framing so they don't curl away from the corner boards after years in the weather. Inside, the drywall must be backed solidly so it doesn't crack in the corner joint. The ends of adjacent walls meeting at the corner must be securely fastened to each other to keep the building solid.

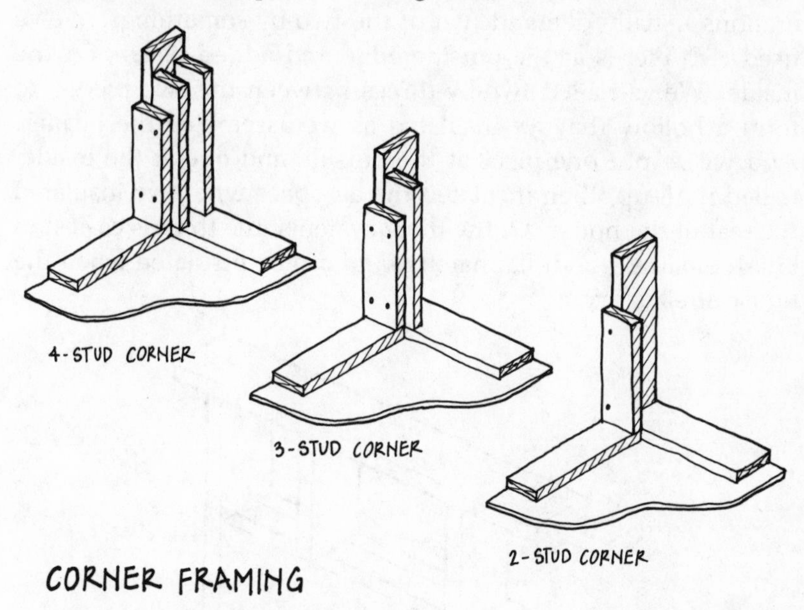

CORNER FRAMING

The illustration shows two-, three-, and four-stud corners, viewed from the top. The two-stud job has plenty of room for insulation and is easy to build, but it is hopelessly inadequate. Only one corner board has good nailing, with nothing at all for the clapboards, and the sheetrock must be attached with metal clips to the adjacent stud, practically guaranteeing a crack. There's no nailing for the baseboard on one wall, and any damage to the nearby sheetrock is hard to repair. It's a dud. The three-stud corner is an improvement for the inside, but lacks good outside nailing. Apple Corps goes all the way with four studs. We leave a pocket for insulation, we have great nailing inside and out, and our electrician can even run wires around the corner without too much trouble. We take care to fasten the single adjacent wall stud often and well to the triple we start with, to keep the corner sheetrock joint together.

The move to two-by-six walls meant a change from the traditional way of building headers. Formerly, a header was two pieces of two-by-something with a piece of ½-inch plywood sandwiched between to make it 3½ inches thick, the width of a two-by-four. A two-by-six is 5½ inches wide, and so must be the header, to provide good nailing inside and out. Strength considerations usually demand two of the two-by-somethings. We've used two pieces at the outside edge and added nailers for the inside. We've nailed two-by-threes between the two pieces, to form a hollow that we insulated as we assembled the header. And we've put one piece at the outside and one at the inside, nailed it all up, then insulated the air space when we insulated the rest of the house. Of the three systems, the third is easiest to build, ensures good, flat nailing, and can be insulated when the house shell is dry.

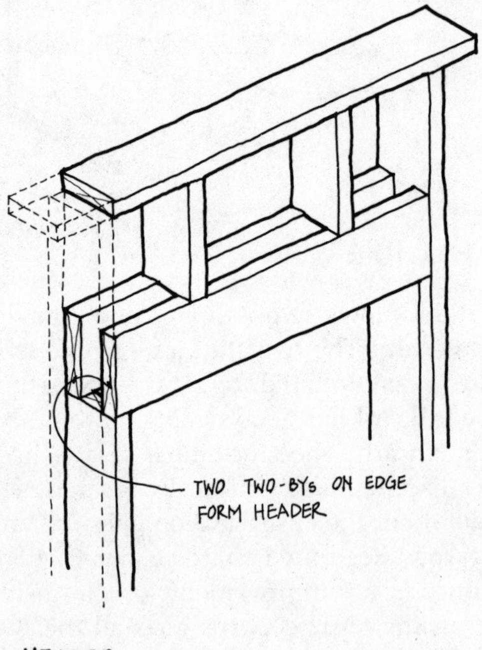

HEADER

If the design allows it, we make the header deep (that is, high) enough to fill the space from the design height of the rough opening to the bottom of the top plate of the wall. The alternative to this full-depth header is a narrower one with short pieces, called cripples, over it, to hold up the plate above. The multiple small pieces are tough to nail accurately and hard to insulate well. If the ceilings are high or the windows low, we need cripples, otherwise we try for the one-piece header. In most cases, this will result in a header that is stronger than necessary, but we don't mind. An especially wide opening — say, for a twelve-foot triple sliding glass door — takes a huge header, usually a double or triple two-by-twelve. We try to avoid such a monster, as it moves around quite a bit from season to season. Instead, we often use a normal-size header over the slider, then beef up the rim joist in the floor above by doubling or tripling it. This leaves us with a reasonable header in the wall and enough total support for the loads above.

Rough window sills are commonly made of a single two-by-six. One is enough support, providing it is a straight and sound piece of lumber. Doubling the sill is a little more work and wood, but stiffens the window opening quite a bit. If we are going to install colonial trim of any serious width at all, we double the sill to provide nailing.

This is a time when you might hear some advice you can best ignore. You're touring your newly framed contemporary with your brother-in-law, who has just had a big custom colonial built. He notices that your carpenters have framed the windows with a single rough sill where his were doubled. Eager to practice his one-upmanship he points out the inferior nature of your sill, and, by extension, your entire project.

Smarting, you must go and talk to your builder. You tell him you're afraid to offend him, but your brother-in-law made some sense, and you can't get it off your mind. Your builder explains that your trim, inside and out, is different from that of a colonial. You don't need something to nail to behind the sheetrock the way your brother-in-law does. And strength is not a concern in this area. In any event, for some builders, as for me, it's simpler to double the rough sills than explain the reason for a single one.

If your builder is confident of his work, he will not mind the occasional questioning of his methods. If your questions are reasonable and not malicious or judgmental, he may welcome the break and the chance to display his knowledge. If you think he ought to change something he's done, tell him so. You'll make trouble for yourself if you persist in asking questions to avoid making a declarative statement. Better to say, "Please put in double sills" than "Are most people's sills doubled?"

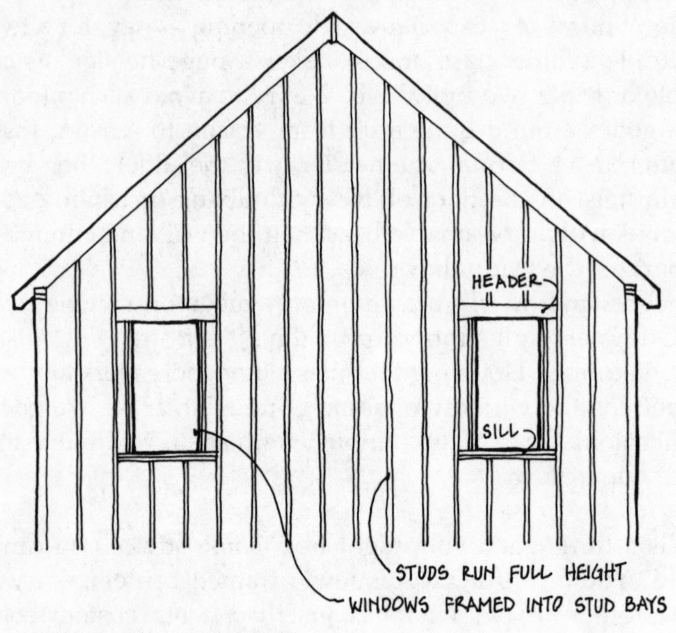

BALLOON FRAMING

There is an exception to the proper symmetry of a rough opening, and that occurs in a balloon-frame wall. Most houses are built using the western or platform-framing method. We build the first-floor deck, then the walls, then the second-floor deck, then the second-story walls sit on that deck, and so on. In a balloon frame, the studs run from the first-floor deck (or even from the foundation sill) all the way to the top of the upper-story walls. The second floor is hung on the sides of the studding. Balloon framing is slower to build than platform framing, mostly because it's easier to rest the deck on the walls than hang it on the inside. In certain places, usually where there isn't a story above, we use the balloon frame to make a stronger wall. A common example of this is a gable-end wall not braced by an attic, where balloon-frame studs are stronger than the two-part platform studding. (By the way, the use of "story" to denote the floor of a house comes from old English castles. Tapestries or painted windows on the stairway landings told tales of the occupants of the castle or of their knights, and walking up the staircase brought you to the next story.)

In balloon framing, the rough sills and headers run between the common on-center studs, and the sides of the rough opening are filled in after that. This makes it unnecessary to run double studs all the way to the top of the wall, which is expensive and clumsy when the wall is so high. A balloon-frame opening doesn't always look symmetrical, but its major failing is in the area of fire safety. The full-height studs mean clear paths for fire to travel swiftly from basement to roof. In the platform frame, each floor interrupts the path of fire up the walls. Building codes require solid wood blocking at every ceiling and floor and at walls parallel to the line of travel of stairways.

We build just about all our walls horizontally on the deck, then tip them up and fasten them in place. It's just easier to work this way than hanging out in midair. We snap lines on the deck to mark the inside (room side) of each exterior wall, making sure they conform to the sizes in the drawings. We lay the top and bottom plates on the deck, and nail through them into the studs placed on the layouts. We frame the openings next, working upward in the wall to make sure everything stays level. We don't force sticks in where they don't quite fit; our goal is a sturdy, flat, smooth, and well-nailed frame. Each joint gets at least two nails, and any two pieces flat against each other get several nails. Using plenty of nails takes more time but makes a sturdier building.

The story goes that a certain sawmill operator was also known for the economy of his building techniques. One of his mill buildings, rudimentary anyway by custom, simply fell down one winter because he didn't use enough nails to hold it together. Common sense was eased aside by thrift, and fell with the structure under the weight of the snow. With this in mind, before I finish framing, I cast a critical eye on what I've done, checking that the studs are straight and even. One reason a good carpenter's work looks effortless to a novice is that the seasoned man keeps the whole job in mind. While the amateur is concentrating on the task at hand, the professional is thinking of the steps ahead. This is a key to good finish work, because the frustration of fastening finish trim, cabinets, or shelves to uneven walls makes crisp work unlikely.

Sheathing

With the wall framing completed and squared, it's time for sheathing. The traditional sheathing for framed buildings was boards, roughly an inch thick, nailed horizontally or diagonally across the studs. When the housing boom after World War II brought us plywood, boards began to disappear as a sheathing material; today they are almost unheard of. Plywood is just plain better. It comes in large pieces, so it goes on quickly and means fewer joints between, so winds are turned away. Plywood makes a sturdy and enduring wall. It holds fasteners, like clapboard nails, pretty well, and is itself easy to nail to the framing. Since it comes from sliced rather than sawed logs, less of the tree goes to waste in sawdust. It's tough to install by yourself on a windy day, but otherwise, it's great for sheathing.

Plywood does have a few drawbacks. In this energy-conscious age, sheathing with more insulating value would be useful. The glue used to stick the plies together is thought to emit dangerous fumes even after it is installed. And some designers think the glue layers constitute a moisture-vapor barrier that traps dampness in the house's insulation. Sometimes, it seems, the more we know about a subject, the less we can be certain of. Except this: any problem plywood visits on us pales before those of flakeboard.

So what about insulating sheathings? They come in sheets, like plywood, and are made from plastic foam or fiber, sometimes covered with foil. None will hold nails; clapboards or other siding must be nailed into framing members, right through the sheathing. They contribute little to the strength of the building; other means of bracing the walls, such as diagonal metal or wood strapping, must be used. The stuff is light to lift, but a handful in a wind. Three quarters of all deaths in building fires are said to come from smoke inhalation, some aggravated by burning plastics with their deadly fumes. Carpeting and upholstery are most often implicated, but foam sheathing in a fire is dangerous. Yet it does insulate.

Carpenters have gotten used to trying new products all the time, but anything that makes a building less sturdy is hard for me to accept. Insulating sheathing makes nailing things onto the outside of houses less secure, especially around openings, where the nailing should be the best. Installing clapboards over, say three-quarter-inch foam requires long, bendy nails and is squishy business; when the nail is given the last tap, it bottoms not on rigid plywood but on soft foam. It's hard to tell when the nail is home and the clapboard (or casing or corner board or frieze) secure. Seasonal movement of the siding causes the nail holes in the foam to become open, cold pathways into the wall. And, of course, the nails may then loosen their hold on the siding — far less likely with a solid nail base like plywood.

Foam sheathing or foam insulation over plywood sheathing has been blamed for a range of siding failures in recent years. It seems the foam, nicely resistant to moisture penetration, causes the siding to buckle and split. Clapboards are especially sensitive to moisture behind them, and suffer the most from foam sheathing. Some recent studies show that foams steadily lose the Freon gas their bubbles were formed with, and with it goes up to 20 percent of their insulating value. And isn't Freon implicated in reductions in valuable atmospheric ozone?

Yet it does insulate. I've used foam sheathing *over* plywood sheathing, and I don't even like that. I won't live in or pay the utility bills for your house, so perhaps I have a skewed vision of what's important. I'd rather see thicker walls with the insulation *in* them than use foam sheathing. I've never built walls thicker than two-by-six and rarely covered them with foam, but higher energy costs in the future may make even our tight houses seem barely insulated. If, right now, I were to build a house for my next forty years, I'd use two-by-eight studs (or two rows of two-by-fours), twenty-four inches on center, and plywood sheathing — no foam at all.

Another product that is called sheathing is building paper or film. In nineteenth and early twentieth century houses with heathing, the paper was used to cover the boards and me of the wind out. Then some builders started using saturated felt (tar paper), normally used under roof shingles, to cover the boards. This waterproof layer kept the rain off the sheathing but caused many problems. Moisture entering the siding through cracks or near flashings couldn't escape to dry out through the back, only straight out through the painted surface. The result in many cases was blistered or peeling paint, and much aluminum siding was sold as a result. Never use felt.

Tar paper has largely been replaced by construction film, the white "paper" you see on houses under construction. The same product is used for the lightweight and very tough envelopes that air express services use. This film stops air from penetrating but allows moisture, in the form of water vapor, to pass through it. Under siding, it keeps cold wind out of the structure but lets the siding breathe, minimizing moisture build-up.

The film suffers a little on installation. Although it comes in a wide roll for good coverage of walls, the roll is heavy, long, and slippery. So it's easy to tear the stuff at the staples until most of it is fastened to the sheathing. Just a little wind makes the job torture; the sheet ends up buckled, torn, and misaligned, so it doesn't work as well. Joints between sheets are supposed to be taped but rarely are. Broad manufacturers' claims of 33 percent reductions in heat flow through a wall are wild, and could only be achieved in a lab house wrapped as for parcel post. Withal, we put up lots of it. I consider it a decent way to cut air infiltration, though I think its wide use stems more from hype than from substance.

There's another way to go, often taken by designers of small or inexpensive commercial buildings. The material is a combination of sheathing and siding in the form of textured plywood, often inclusively called by the name of one of its variants, T 1-11 (for Texture One Eleven). It is available in several designs, and sheets come up to twelve feet long. Since sheathing and finish siding are applied in one operation, it saves on labor costs. T 1-11 is at its best on long blank walls under wide roof overhangs. Fitting it around openings is tricky and time consuming and never quite right. Some builders have had problems with the layers delaminating, or peeling off, because of poor gluing, so keeping it relatively dry is a boon. Any hole or gap in the siding means an infiltration path right to the insulation, so careful fitting, particularly around openings, is a must.

It does save money, though, and I've used it on a house where the young couple planned to side the building with clapboards when they could afford to, installing them over the T 1-11. This seems an intelligent use of a somewhat marginal material. As a long-run choice on a house it is poor, because it just doesn't age gracefully; it always looks and acts like the low-dollar product it is.

The conclusion on sheathing, then, is: use plywood. In rare cases, T 1-11 makes sense, and foams bring their own problems. The sheathing is really the shell of the house, and our goal is to make that shell rugged enough to last a century or two. Remember, I'm asking you to pay for a building you will never use up.

Now that you know which sheathing to use, let's see how it gets attached. The framed wall is lying on the deck, square and straight. All the studs and headers and jacks and sills and cripples and blocking are nailed tightly.

We're going to install the plywood horizontally, so the long edge is parallel to the deck. The lowest course of wall plywood should also cover the mud sill and the edge of the deck, so we measure the thickness of all that; it's usually right around a foot, which is the amount the plywood should hang below the bottom of the wall. We nail the plywood with sixpenny hot-dipped galvanized commons, sturdy little nails with good holding power.

We like to cut the plywood around the window openings before nailing it on. Another method is to cover the openings completely, then cut out the windows and doors from inside after the wall is tipped up into place. Either works, though the first is quicker for us and leaves neater and less splintery scraps. In any case, the nails go eight inches apart, except around the openings, where they should be five or six inches apart. Nailing the sheathing securely means a well-braced frame and a consistent wall thickness. (Loose sheathing, with a gap between it and the frame, means a bulging wall and spells trouble in finishing around the openings.) As a rule, the ends of plywood sheets in adjacent rows shouldn't land on the same stud.

After all the sheathing is secure, we tip the walls up and nail them into place. This takes roughly two men for every ten feet of wall, and we often combine our crews for those brief, glorious moments. We brace the standing wall to the deck with long two-by-fours every eight to ten feet. These braces are terribly in the way but terribly necessary. These walls are much like sails, and a storm warning heightens our senses much as it does those of ocean-going yachtsmen. When all the walls are up, we'll adjust the braces to make the walls perfectly straight before we continue with decks or roofs above. We generally leave most of the wall braces in place until the plywood is all nailed onto the roof; that's when the house is fully framed, and as strong as it will get. Removing the bracing is exhilarating, the first time we get to see the enclosed space clean and unencumbered.

Interior framing

Then, of course, we fill it up again with the partition framing. We move the braces temporarily, then nail them back up to or through the partitions. Many builders frame the partitions as they finish the outside walls of each story. Others wait until the roof is on so they can frame on rainy days. Either system is fine, though bearing partitions (those that carry the weight of the floor above) must go in with outside walls.

We have taken care, when framing the outside walls, to provide attachment points for the partitions. Before energy awareness was current, we framed partition posts, three studs in a U shape at each partition-wall intersection. These made for great nailing for everything, but left an uninsulated cavity at each intersection. Now we nail a two-by-eight flat onto the end partition stud, or into the outside wall, as we're framing. This allows the insulation to continue uninterrupted past the partition, and uses less lumber to boot.

Virtually all house partitions are framed with two-by-fours spaced sixteen inches on center. Openings for doors, archways, pass-throughs, or medicine cabinets are surrounded with the same stuff. Doors and other large openings require two studs around them for stiffness and to provide nailing for the trim they'll ultimately get. Interior doors have rough-opening sizes just like their outer-wall counterparts. Headers are simpler, usu-

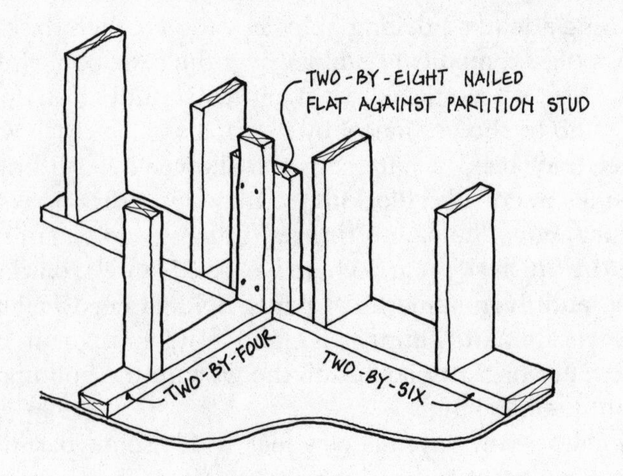

PARTITION NAILER

ally two two-by-fours laid flat and stacked up over the opening, except in bearing walls. The key to a good job is care in choosing the straightest studs for the openings, nailing everything tightly, and keeping all the framing smooth and even.

Blocking deserves mention here. Blocking provides nailing (or screwing) attachment points for finish materials and fixtures. Anywhere you intend to fasten something permanently to a wall or ceiling should have blocking framed in. Bath accessories, all cabinets, fiberglass showers, built-ins, staircase trim, and handrails need blocking. If we're not sure exactly where this stuff is to go, we might block the whole area with plywood to be sure we're covered. Toggle bolts are a pain, molly anchors rarely are strong enough, and blocking is easy to do if the architect and builder think of it beforehand.

For kitchen cabinets we usually nail in solid blocking two inches below and four inches above the counter height by installing a two-by-six flat into the wall. This piece should be of full length and let in to the studs, which means the studs are notched out so everything is flush (even) on the surface. This gives us a chance to straighten out crooked studs, and provides solid nailing for cabinets and backsplash. Being able to fasten things securely at the joint where counter and backsplash meet helps keep finish surfaces tight over the life of the installation.

Planning adequate nailing helps in various other places. If you want a recessed medicine cabinet over the bathroom sink, make sure the studs are spaced for it. Studs should clear the center line of a tub or shower faucet by five or six inches each way. The plumber may need a nailer for the shower head fitting. Wall-hung sinks need solid blocking so they don't droop. We fill the walls just above the stair stringers (framing), to be sure we can nail on the finish skirts anywhere we like. Mantels, outdoor deck railings, and even some toilet paper holders need nailers. And most vertical-board sidings demand solid horizontal blocking every twenty or thirty inches all the way up the building! Don't leave out the blocking.

In houses over, say, sixteen feet wide, some partitions are used to carry the weight of the second floor. These bearing partitions are built along with the outside walls, are structurally tied to them, and brace them. (That's why in remodeling we're cautious about taking down bearing walls.) We ensure ourselves a level second floor by cutting the partition studs from the same pattern as the exterior-wall studs and by using the same double plates on top to tie the whole together.

Partitions should be designed for what will be buried in them. In a house heated by warm air the partition framing should accommodate the ductwork. In any house, pipes must have room to go through the partitions. Careful planning for plumbing runs will save expensive ad hoc deliberations over where the pipes should go. Special avenues for plumbing and HVAC (the trade's acronym for heating, ventilating, and air conditioning) runs, called chases, are often built into odd corners.

Another consideration in the framing of partitions is soundproofing. You might want to isolate some rooms from others, or one room from all others. Sound moves chiefly through air, and tight-fitting doors are the first line of defense. Sound waves also vibrate building materials, which transmit the sound to adjoining areas. Two sets of studs, offset, one for each side's drywall, help reduce the transmittal, as does a double layer of drywall, which tends to absorb more of the vibrations. (Note that offset studs won't restrict a fire to one cavity in a wall, and may allow it to spread much faster.) Fiberglass building insulation between the studs and the joists helps, as does using solid, not hollow, doors — both of which forestall fires, too. Drywall can be attached to flexible metal Z-channels, which absorb some of the vibrations. The quietest room would have all this stuff, plus deep carpets and soft wall coverings. What you'll probably end up with is wall and floor insulation, a tight, solid door, and maybe double drywall on the noisy side, an effective and not too expensive compromise.

Second floor

The second-floor deck isn't quite the same as the first. It supports less activity and fewer concentrated loads (except for bathtubs). Its structure and mechanical systems are hidden and hard to change (the "guts" of the first floor are usually accessible from the cellar). You may want to soundproof the upper floor or heat it separately. And the second floor is often framed without girders; in most cases, the many first-floor walls can support it.

It also requires an expensive finished staircase. The stairway should be figured first, even though it's built after the second floor is in place. The stairs must be safe, good looking, comfortable, and get you where you want to go. They are expensive and deserve your careful consideration. I'll talk about their finished qualities in chapter 14; this section will cover location, safety, and ease of use.

Stairs are often placed in the front hall, the focus of those who use the front door. That makes sense — they cost a lot and look nice, so show them off. They must, though, lead to and from the right places in the house. There's a lot of traffic around stairs, so they should end up in a buffer zone between travel and activity. A hallway or foyer keeps kids flying downstairs from crash-landing on the cook.

As houses get smaller, stair layouts get more compromised. Stairs take about 10 feet by 3½ feet out of the downstairs floor plan, and slightly less upstairs, in a house with 8-foot ceilings. A stairway that doubles back at a landing takes, at minimum, a 7-foot square. If either rectangle won't fit on your floor plan and leave you room to get on and off the stairs at the top and bottom, you have a problem. It's not so good if your stairs deposit you facing a wall, forcing you to turn abruptly. Placing the stairs in the plan is tough, and should be considered quite early, before room layouts are decided on.

Picking a central spot in the house makes sense. You shouldn't have to walk too far to the stairs from any downstairs room. You don't want to go through other rooms to get to the stairs, certainly not too many. Especially when it's time to replace the box spring in the master bedroom, you'd love to be able to go from upstairs directly outside. Dragging your floppy old mattress the length of the living room is the stuff of Saturday morning cartoons, complete with sound effects.

Stairways also act as a ventilation shaft. Air is exchanged efficiently between any two areas of differing temperature. An open staircase siphons warm ceiling-level air from the downstairs and replaces it with a chill breeze descending the stairs themselves. If your stairs end up in the living room and you're sitting near them, you will know this firsthand. Even with forced-warm-air heat this flow is hard to control. So keep the stairs away from sitting areas or where someone might want to lie on the rug.

Natural light on the stairs helps show them off and makes them safer, so they should go near a window or glazed door. A skylight over the stairs will offer the same effect, but the cool downdraft will be worse. In any event, make sure there is enough artificial light to keep the whole flight of stairs out of dark shadow.

People are accustomed to the thirteen or fourteen steps in a standard house staircase, and more than that gets tiring. The number is a function of the ceiling height and the steepness of the stairs, and varies only by one or two in most cases. The many scenes in old movies shot on endless grand staircases were likely there because the actors were too tired to make the whole flight at once. Your house design probably won't call for more than fourteen steps, but if it does, you may consider going upstairs a career.

Most building codes mandate that stairs used to leave the house in an emergency be at least thirty-six inches wide. This is a comfortable width anyway, though tight for two people passing. A bigger or grander house should have a wider stair; four or six inches make a big difference. A staircase enclosed on both sides by walls is rarely used in a fancy house, but it needs the extra width even more, to make it commodious.

Codes also demand a minimum headroom of about eighty inches, the height of a standard door. Headroom is measured vertically from the front (nosing) of any tread to the ceiling area over it. Often, making the headroom code-legal makes the rooms above the stairway smaller. Many house designs, in trying to squeeze the floor plan for its last few square inches, cheat on this dimension. Most people duck their way downstairs without a second thought, so the cheating is tolerable. In an emergency, though, you don't want the tall fireman carrying you down the stairs to knock himself out.

Other critical inches in the staircase design are the rise and run. Rise means the vertical distance from the top of one tread to the top of the next. Run is the horizontal distance from one riser to the next, risers being the boards your toes graze on your way up the stairs. Stairs are easiest to travel when two rises plus one run add up to about 25 inches. The rise and run are planned in relation to each other. It's best not to have the rise higher than 8 inches or much lower than 7, with 7½ a good compromise. This puts the run, from the formula above, at about 10 inches, which is fine.

Outdoor steps sometimes have rises down to four inches with very long runs; these work but are quite uncomfortable to use, and are unacceptable indoors. The problem with oddball runs and rises is that we're conditioned and expect to step up a certain amount when climbing stairs. The steep rises and short runs of antique houses are interesting to tourists but tiring in everyday use. And the glamorous wide treads and low risers of cinema mansions make everyone, not just John Wayne, want to skip every other step.

Run and rise calculations should take into account all the steps, including bottom, top, and landings. If there is to be a carpet runner attached, the builder must know ahead of time; the first riser will have to be shorter, and the last higher. Your legs get used to the rise by the second step up or down, and all steps must be the same to keep you from stumbling. A ⅛-inch variation in rises is enough to trip someone who is in a hurry, and a staircase is a bad place for a trip.

To save space or accommodate a specific floor plan, architects sometimes design a turn or two into the staircase. Turns come in several versions, from a simple landing turn to a spiral stairway — all turn. For safety, any departure from a straight staircase requires careful attention to detail in design and construction.

The landing turn is easy. The designer saves length in the stairwell by turning the bottom, or less often the top, to the side. In most cases, you'll start up two or three steps to a landing, turn, and continue straight the rest of the way. This design allows a building under thirty feet wide to have a stairway ending at an upstairs hall, and leaves space for good-sized rooms on the other side of the hall. This design takes up more room *along* the house on both floors, and makes the stair hall squarer instead of longer.

A variation of the landing turn is the winder. In this setup the stairs turn a corner with wedge-shaped treads that taper smaller toward their pivot. Winders look great but are tough to build and walk. Codes require a minimum of nine inches of tread width one foot from the narrow side, as well as certain minimum widths for that side. In practice, we've built winders that taper to just about nothing without violating the nine-inches-at-one-foot rule. Because people tend to travel the shortest path around a corner, where a winder's treads are narrowest, the traveler must decide at each step where each foot falls. This may be an intellectual and physical exercise best practiced elsewhere. In short, winders are pretty but inherently unsafe.

A bastard version of the winder is the split landing. In this style the landing is divided on the diagonal, forming two right triangles with the step at the hypotenuses. The designer gains thereby one extra riser and a little more floor space. Traveling one of these oddballs isn't bad, because the triangular tread widens quickly to a safe width. One problem with the split landing is integrating its style into the house's; it seems to look best in a contemporary, out of place in a traditional house. A second difficulty is finishing the underside, which is neither smooth nor regular, and worse than a winder's. It's a good way to go when space is tight and the style amenable.

The elegant curved staircases in grand houses are very expensive, and seldom used by anyone with a budget. More likely to appear in the specifications, though, is the spiral stair. I hate them. They are hard to walk up or down, often narrow and possessed of elastic railings, and impossible for moving furniture on. I've even gotten so I don't like the looks of the darn things, though I used to. They save some space, true, but at what a price! I might be convinced to put one in sometime, but I'll leave no other option unchecked. Really, the only virtue I can see in a spiral staircase is that some dogs won't climb them.

The framing of stairs must be safe, solid, and simple. Safe means, normally, three two-by-twelve stringers (notched supports for treads and risers) for up to forty-inch-wide stairs, four stringers for up to fifty-inch widths. Stringers must be solid lumber, not two-by-sixes with triangular blocks nailed on. Stringers on each side must be nailed to the wall framing or doubled. Stringers over about ten feet long will be springy unless doubled.

Where stringers attach to first and second floors (or landings) is a common weak point, for a simple reason. If second-floor joists are two-by-eights or even two-by-tens, and the stairs call for a 7½-inch rise, only an inch or two, or less, of the stringer bears on (that is, is in contact with) the framing. This weakness can be cured many ways, but it must be cured. Wood-to-wood joints and construction adhesive are my favorite, along with adding nailing ledgers to the floor framing.

With the stairs figured, it's time to frame their destination. The second-floor joists may well be the same size as those on the first floor, depending on where the first-floor partitions are. If there is a question about the sturdiness of the floor, I opt for the larger of the choices debated. No one wants a springy floor, and my experience with some old houses and their underframed floors makes me conservative. The fine, sturdy old houses I much admire were probably considered overbuilt when new. I hope to leave that kind of legacy.

So I double up the joists under the partitions parallel to them. I install bridging, wood only, if the spans warrant it. Metal bridging is particularly insidious in a second floor, where nothing can be done to silence its thrumming. I negotiate with subcontractors they make any framing cuts. In their zeal to stretch their or ducts from here to there they sometimes nick, notch, or eliminate important joists or studs.

von't cantilever any house framing outside; that's a guaranteed trouble spot in a few years. "Cantilevering" means letting framing members stick through the side of a structure to hold up something — a balcony, a roof overhang, or a yard light. You can be easily seduced by the possibility of having a little sundeck off your bedroom when all you have to do is run the floor joists out a little longer and put in a platform and railing. In theory, using pressure-treated everything, you might build a lasting deck. In real life, most anything cantilevered outside eventually leaks water into the main structure. It's tough to fix these leaks, so I reject them at the design stage. Any outside structure attached to the house should be built as an add-on, not tied into the innards of the house.

We're careful when we build the second floor. Before we nail down the plywood we make sure the joists are all straight (this is called "on layout") and nailed everywhere they touch the walls below. We glue sheathing down with construction adhesive and use galvanized nails to fasten it. We crown every joist and header, just as we did on the first floor, so years of use will leave a true structure. We want sturdy, quiet, and stiff floors, and these are the ways to get them.

More walls

The second-floor walls come next. The deck is clear, so we take this opportunity to check the dimensions and squareness of the building. Sometimes the deck is slightly off size, and we can get it perfect again by snapping chalk lines on the deck, marking the designed size. That way, it's easy to hide a quarter-inch or even a half-inch discrepancy caused by rain swelling or the dread human error. Taking time now to check the size means a straighter roof; once the second-floor walls are up, they define the size, and if they're off, they're off for good.

In a conventional two-story house, the second-floor walls are much like those on floor one. The windows are framed the same way, though we sometimes must position them just under the overhang of the roof. In general, I like to see second-floor win-dows centered over those below. Lining them up seems to appeal to the eye, to make sense to the viewer somehow. Maybe it's just that misaligned windows don't look right unless some other feature of the design makes the placement seem logical to us. Windows scattered around whimsically (by intent or inattention) force us subconsciously to justify their appearance or try to ignore it.

Many contemporary house designs employ odd-height walls, deviating from standard eight-foot ceilings. If these walls are shorter than roughly four feet, we balloon-frame the first- and second-floor walls as one unit. The short second-floor wall draws its strength from the taller first-floor section, braced by the second floor itself. As long as it doesn't extend too far above the second floor, this upper wall can handle some of the outward thrust of a gable roof.

It's useful to know some of the stresses imposed on buildings by their roofs. Bear with me as I digress into the realm of roof forces, so you can learn what work your walls must do. You may want to refer to the illustration on page 141.

In any roof, the dead load (the weight of the structure and the shingles) and some of the live loads (rain, snow, wind, and the chimney sweep) push down on the roof's supports. Common gable roof rafters, the regular, angled beams the roof surface rests on, push *out* on their bottoms as well as straight down. (Pinch two pencils together in a wigwam shape, rest their points on the table, and push down. The down force translates into a spreading-apart force at the bottom of the "roof.") In a standard house with flat ceilings, the ceiling joists tie the bottoms of the rafters (pencil points) together. Without the flat-ceiling framing, a gable roof will push out its supporting walls and fall down.

In rooms with partly sloped ceilings, the spreading can be limited by tying the rafters together partway up with horizontal pieces called collar ties. You may want them high up to get the most headroom, but the lower they are on the rafters, the better. It's possible to make the roof sturdy even with high-up collar ties by using oversized rafters. Heavy rafters, often two-by-twelves, won't bend much from the ties down, and therefore won't push ir bottoms. The joints and sizes of all the pieces make this system work.

thedral, or sloped, ceilings use a ridge beam to 'e of the roof. With a cathedral ceiling or shed and most live loads (wind and earthquakes ex- must be carried straight down by gravity. A ridge beam makes each half of a cathedral roof act like a separate shed roof, pushing down and not out. The ridge beam must be strong enough to carry this weight, and that means *huge* if it is longer than twelve feet or so. If you see a house with a cathedral ceiling in a room wider than eight feet, and you don't see a big ridge beam in the framing, find out why not. I've been asked to install ridge beams in finished houses built without them, so I know builders sometimes leave them out.

The ends of the beam must be supported by posts capable of carrying the roof's loads. These posts, if they are to be hidden, are normally made up of studs nailed together. They should rest, eventually, on the foundation sill, or on something of like size that rests on the sill. Beware, for instance, of a design that calls for a ridge beam carried on a post bottoming in the middle of a header for a sliding glass door. Understand, the house may be fine for a few years. When the sag in the header makes the slider into a stationary door, though, it's too late for simple corrections.

There are a few other things you should know about upstairs walls. Tall gable walls — say, in a cathedral-ceilinged room with no second floor to brace the walls — should normally be framed with full top-to-bottom studs rather than be divided halfway up by horizontal members. A really big wall may even need two-by-eight studs to be stiff enough in a gale. Gable walls in a cathedral-ceilinged room should run up to the top of the end rafter, not under it, so the insulation can completely envelop the house. Again, the simplest yardstick of good framing is neatness, with no gaping voids at joints, few irregular sequences, and few oddball pieces stuck in here and there.

Some house designs, notably Capes, call for knee walls, short upstairs walls parallel to the ridge that run up to the slope of the roof. Knee walls work best if framed as regular walls with top and bottom plates, the top plate notched into the rafters, straightening them. It's also possible to run knee wall studs

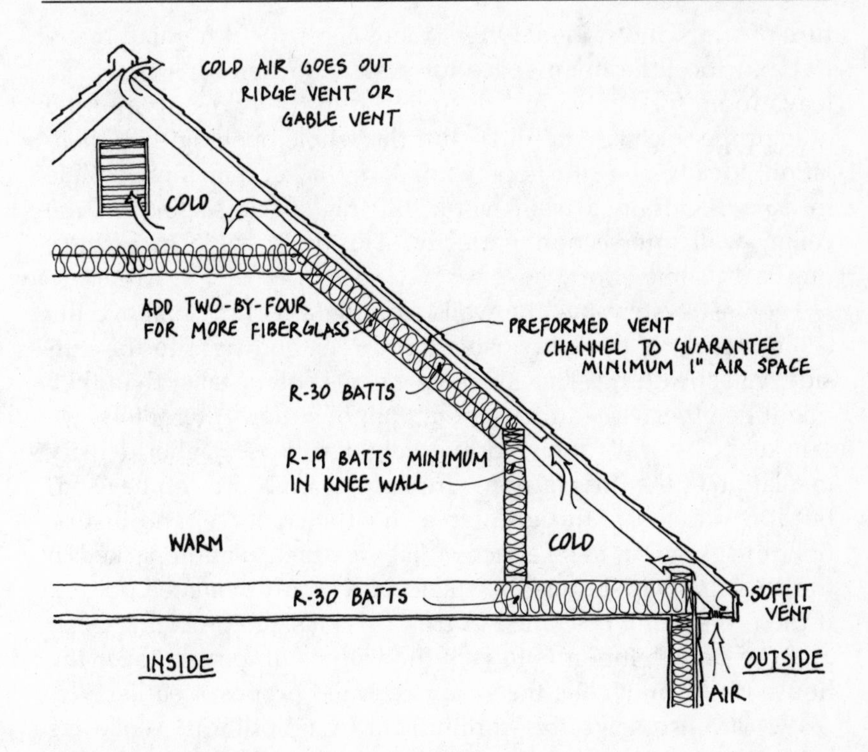

VENTILATION

alongside rafters and skip the top plate, but this makes for poor sheetrock joints. In any case, it's best if floor joists, knee wall studs, and rafters line up.

Our other concern as builders is the sloped part of the ceiling, the part that follows the roof framing. Capes can often use fairly light rafters, like two-by-sixes or two-by-eights. These are strong enough in a good design, but too shallow for adequate insulation where they are exposed in a heated or cooled room. One solution to this dilemma is to use two-by-twelve rafters and nine-inch fiberglass insulation. On a partly sloped ceiling, we can add two-by-fours on edge to the rafter bottoms at the slopes only. Either system will make up the R-value we like to see in a ceiling (at least R-30; R is resistance to heat flow). Both, of course, cut down on headroom.

One fairly weak point in many houses' insulation is the juncture of walls and rafters. Often, there is plenty of framing material and too little open space for insulation here. It's crucial to leave room for ventilating air to get from the eaves up into the attic proper, so we can't just stuff the whole opening with insulation. Ideally, the attic floor insulation, thicker than any possible in a regular framed wall, would fill this space, sealing off the ceiling-wall intersection perfectly. This is possible in practice only with some effort.

The best system has the walls running a foot or so above the ceiling level, so the attic insulation can extend right to the outside wall sheathing. This means the walls will be taller than they might be otherwise, and aesthetics might suffer. Alternately, we can keep the walls their normal height and use higher-density insulation at the intersection. Urethane foam board can be used, but it is difficult to fit tightly in a constricted space, and insulation must be tight to be effective. Fiberglass insulation packed in there would help, but would still leave a poorly insulated pocket. If the roof-ceiling is formed of trusses, the same problems exist, and the only solution is to jack the trusses up on legs over the house walls, much like the extended wall I proposed earlier.

We can also solve this conundrum by ignoring it. While we know insulation is very important, spending hundreds of dollars to solve a few dollars' worth of problem just isn't practical. Even with inexorably rising energy costs, the benefits of going through a major operation to slightly reduce heat loss in one narrow band of the house seems irresponsible. Tight construction methods and materials and careful insulating of the whole house ensure the best insulation value.

Second-floor partitions don't have much to hold up unless the design calls for a usable attic, a rarity these days. When an attic or storage space is called for, the second-floor ceiling is made just like a floor, and the joists require a bearing wall. If you don't plan to have a usable attic, roof trusses are the alternative. Trusses are small pieces of lumber assembled into units that form the structure of the roof and ceiling. They give some flexibility to the designer; roof trusses rest only on the building's outside walls, so they need no interior supporting walls below, as a site-built roof structure usually does.

The trouble is, trusses are seldom built well enough. One of

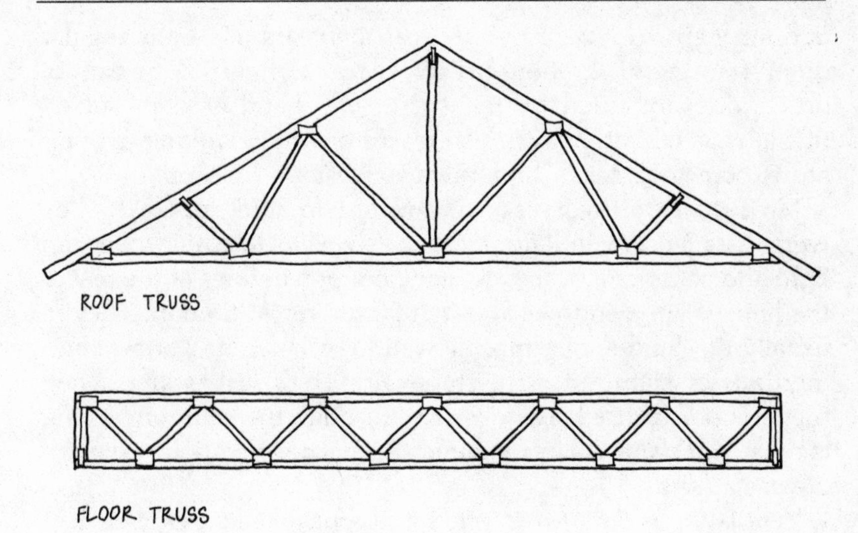

TRUSSES

the economies of truss manufacturing is being able to use poorer grade lumber than would be acceptable in solid-wood rafters or joists. This is great for the price, but dooms the truss to a second-rate existence. For the trusses are sometimes unimaginably more crooked than the solid wood they replace. The cheaper lumber they're manufactured from kills the chances of their being straight, even as it offsets the cost of the labor needed to build them. We use trusses where economy is a main concern, and rarely elsewhere. In addition, roof trusses, with their webs of two-by-fours, make an attic worthless for storage.

The attic

Attics are losing the battle of the square foot, the struggle to make every expensive one of them livable. Floor plans and section drawings are assaulted for the possibilities of tenancy from the first budget meetings. In this war the attic is a civilian casualty, passive space inviting conquest. You can, however, end up the winner on the plans and the loser at home.

Attics were great for storage, and they supported extended families with their generations of furniture. Attics have thrilled and terrified many youngsters over the years. Here the regular dusty framing of the house displayed a stranger's definition of order, one not ruled by the swift, sure hand of domesticity. Illusions aside, attics were also places of wildly varying temperature, occasional leaks, and, always, wasps.

Most elderly attics have a window or two, rarely opened. They were installed to ventilate the attic as well as to admit their dim light and complement the arrangement of windows in the rest of the house. The windows aren't great for vents, though. They're usually not high enough in the wall to exhaust the hottest summertime or dampest wintertime air. They leak water when they're open. Since water dripping through the bedroom ceiling is more noticeable than the effects of poor ventilation, the windows stay shut.

Ventilation is *the* issue in attics and roofs of all types. The goal of ventilating is to insulate the house proper from the attic and keep the attic in equilibrium with the outdoor air. In practice, this dictates letting as much air as possible move through the attic while keeping rain, snow, and wasps out. I'll discuss ventilation more thoroughly in chapter 8 and stick to attics here.

You'll want to have ten to fifteen inches of insulation in the ceiling below. If you intend to use the attic for storing anything, you must build the floor higher than the insulation. Two-by-six or two-by-eight joists are strong enough for light storage. Another layer of the same two-bys toenailed right on top makes a deep cavity for insulation. I'll nail the second layer crosswise to the first when I use batt insulation, or directly on top for blown insulation. I can't remember when I last built a completely floored attic; people usually end up with just a few sheets of plywood down the middle.

A folding attic staircase is a poor way of getting up. These things are great for leaving the story below free of the encumbrance of a permanent staircase, but they are terrible to climb, especially if you're carrying anything. The best folding ones are none too strong, and are narrow and steep to boot. Roto makes a good metal sliding staircase, nicely designed and sturdy, but the price is high and it's still narrow and steep. If you must use a folding staircase, you can specify a special-order model that is wide and sturdy. Do, or you'll curse the thing forever.

You should opt for a regular staircase if you have the room. Anyone looking for more room in a house he buys certainly welcomes the regular attic stairway, as it leads to a fairly easily remodeled space already enclosed. A design incorporating an unfinished but exploitable attic may give you the space you'll eventually need without killing your current budget. Leave an easy way to get bulky materials up there, though (plywood and drywall can be boomed through a large window), or the job will be daunting and expensive. How you design the attic will make or break the conversion, so make it strong enough and ventilate it well.

8

On to the Roof

THE ROOF IS the essential element of shelter and does the most work for the dollar of any part of the house. It protects the house, its occupants, and their possessions from rain, snow, and sun. It also is the major feature you think of when you picture a house's design. (In current realty parlance, a Cape can be a ranch house with a steep roof.) Our interest is in building a roof that is good looking, functions well, and lasts a long time.

Roofs are designed in a huge range of styles, from the New England church steeple to the sod of the earth-sheltered house, from saw-tooth industrial to the Sydney Opera House. North American house roofs have become fairly standardized in construction and design. The double-pitched (sloped) fiberglass shingle roof works well in most climates, is cheap, and can be installed leak-free with minimum attention to details. I'll examine different styles first, then address the methods of construction.

The most common roof style is the gable, which slopes down on two sides from a central ridge. The pitch of gable roofs can be shallow, like a low-dollar ranch or modular house roof, or steep, like an A-frame. Gable roofs can thoroughly overhang the building in a chalet, or be close-cropped in a development Cape. They

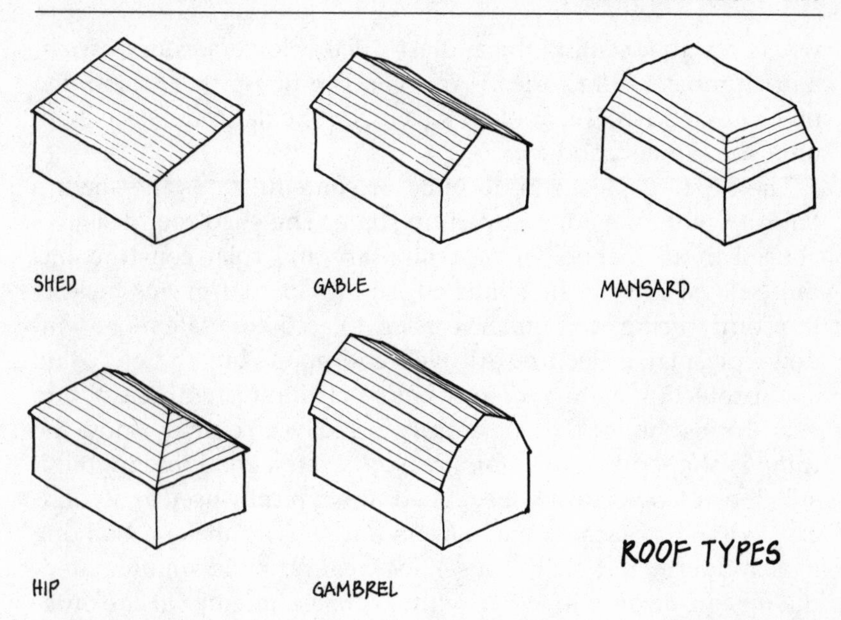

may be littered with dormers in an Ivy League dormitory, or plain and unbroken in a tobacco barn. They may be surfaced with tin, shingles, thatch, tiles, slates, boards, leaves, or photovoltaic cell arrays. This prevalent roof shape is endlessly various and endlessly repeated worldwide. When you think of the ways the gable roof works, it makes good sense.

Think of a basic building, a one-room square. Water runs downhill, so it's only logical to tip the roof so it runs off. Tipping the whole roof in one plane works, but makes for long roof beams and exposes the high wall to more of malevolent Nature. A gable roof, with the center pushed up and parallel with the longer walls of the building, is pretty easy to construct. And the building is made more spacious inside; you can use the area under the roof.

A flat roof will work for most purposes, but is difficult to seal against rain and snow, especially when it's made from indigenous materials. Too, for a flat roof the roof beams must reach from wall to wall in one stretch. That means making them strong and heavy, and therefore hard to come by. A flat roof covers the most building with the least waterproofing materials, though, so it's universal in industrial buildings. (It's easy to walk around on when you're installing the endless ducts, blowers, compressors, and chimneys of factories.) And you can lie up there in suntan oil or in ambush. Technology has made the flat roof feasible but only sometimes good looking.

The next-simplest roof to build or contemplate is the shed, a flat roof with one edge tipped up some. The shed roof design is abused in all manner of contemporary and solar constructions these days. Lifting the south edge of a flat roof means making the south-facing wall under it taller, to accommodate more windows or solar collectors. All well and good, but the enclosing and protecting nature of the house is diminished. It's a rare shed-roofed house that looks truly attractive. You may know the thing is shaped that way for a reason, even a good reason, but it still doesn't look like a house. Sheds are typically used as secondary roofs, on dormers and towers and ells, to make a building look modern. But such houses look choppy and unintegrated, the design thrown together with a careless attempt at the striking. Fifteen years ago I built my first and last.

There are a few decent spots for a shed roof. One is as a large dormer, that source of usable second-floor square footage in an otherwise too small house. A shed roof is about all you can use there, and looks all right as long as it doesn't extend all the way to the end of the building. This is a theme often massacred; make sure you see many elevations of your house before approving a shed dormer. Sheds are used for small dormers, too, though usually less successfully. A shed dormer is easy to frame and trim, is thus cheap, and often looks it. There are exceptions, but in general a small shed dormer should be reserved for a shed.

Another kind of roof that is often misused is the gambrel, in which the straight leg of the gable rafter is bent at its knee. The bend expands the space just under the roof, making more room for hay or for walking around. The bend can be near the top of the roof, a sort of hat-on-a-gable form. Sometimes the rafters are bent far from the ridge and make the roof look like walls-plus-gable-roof, a whole extra story. The extreme version of the latter design often includes windows in this wall-roof section that may be protected by tiny dormer roofs. I much prefer the former style to the latter, and find few gambrel designs that really embody pleasing lines. The good-looking exceptions are stunning, but must be hard to achieve, since they're so rare.

The hip roof and its cousin, the pyramid, slope on all sides toward the top. The pyramid slopes to a point, and works only on a square house, while the hip roof has a true ridge. The eaves of the roof are usually the same height all around the house. This roof is very popular in southern climes, as it is easy to make a substantial overhang all around to shade the walls and windows. Taste and expectations are probably the major factors in a design's regional popularity, though. Technology allows us to bend many house designs to our will, mating them to alien climates in ways our forefathers would never have abided.

The one difficulty with hip roofs is venting them high up, because vent openings must be cut through the roof itself. Roof fans are noisy and use power, and it's hard to get enough passive vents in a hip roof without its looking like a porcupine. Small gables fashioned at the ends of the ridge to contain screened louvers will work, but may not fit with your design. Be sure your roof gets enough ventilation; see the next section for some standards. Hip roofs have simpler cornices (roof trim) than most others, saving some labor and materials, although they are somewhat tougher to frame and shingle than gables. In the end, style is the main determinant of your choice.

The least appealing roof, in its modern as well as in many of its traditional forms, is the mansard. This roof combines a flat or nearly flat hipped upper section with an almost vertical lower section, sometimes pierced by windows. The mansard roofs on the seventeenth-century chateaux of France, carefully integrated into the buildings and with the patina of centuries of history, are quite handsome. Most modern attempts come off less well, and this style has reached its nadir in the fast-food joints of America. This latest version serves to dress up a boring flat roof, hide the ductwork and air conditioners, and provide a surface upon which to hang signs. I can't think of an unhappier ending for a once noble form, and I certainly wouldn't nail one on a house.

Designers often feel that the shape of the house should be able to stand on its own, without embellishment. Why build a Cape with big dormers on both sides instead of a real two-story house if you clearly need that much space? Complexity is expensive, in both initial cost and maintenance, so why disfigure a pure form? On the other hand, a series of rooms with separate roof lines and dormers looks good, even elegant. So it's a trade-off. You can pay the architect to draw you a house with a straightforward design, true unto itself, or you can layer-on roof lines and set-backs yourself with a bought design.

In snow country, a flat roof is a poor choice. Most flat roofs are not truly flat, but tip slightly so water will run off. Snow and ice often block the way, so meltwater stands in puddles, inevitably inviting itself inside. Flat roofs are difficult to ventilate properly, since they have no height to induce natural air flow. If everything works right, a flat roof can be leak-free for years, but a sloped roof is less demanding. Repairs can be made by almost anyone, and reroofing and alterations (like sticking an added chimney through) are much simpler. House builders just don't like flat roofs any more than they like *anything* that leaks.

The cornice plays a major part in the design and function of the roof. Broad overhangs shade windows, and skimpy ones evoke, for some people, our colonial roots. Elaborate embellishments duded up the Victorians' gables and can do the same for yours. Built-in gutters or layers of moldings can add traditional grace to a pleasant design. Soaring stark lines add drama to contemporaries.

You won't have many choices about your cornice because its design is integral to the house's. Specify that it be built from good lumber of a clear grade. Pine is good, and must be kept well painted. Redwood and cedar are long lasting but still need the regular oiling of a stain or paint job. Incidentally, hardly anyone uses hardwood exterior trim, because it reacts more violently to wetting-drying cycles. It's great if you get some overhang on eaves and ends in your design. Keeping the drip lines inches away from the windows and siding will add years to their life.

Before you approve the design of a cornice, you must embrace or reject gutters. Gutters are great when and where they work. They protect the sides of the house and the doors and windows from sheets of rainwater. They gather that water and channel it away from the foundation, helping keep the cellar dry. They fill up with leaves and get plugged up and must be cleaned out, usually from the top of a homeowner's shaky ladder. They contribute mightily to ice dams and winter roof leaks in the snow belt. And they fail noisily in severe winters, dangling by a strap or two after succumbing to their hundredweights of built-up ice.

I will hazard a rule that if your area harbors a foot or more of snow on the ground for six weeks or more in an average year, don't get gutters. In hill country, local climates can be wildly different scant miles apart; you'll have to know what to expect at the exact site you'll build on. Snow and ice problems on roofs are most severe in the earlier months of winter when the weak sun makes few inroads on the icy eaves. (I'll discuss roof icing and its cures shortly, and gutter choices in chapter 17.)

One design pitfall to watch for is when one large roof dumps water onto a roof below. A worse dumper yet is a valley, an intersection of two roofs where rainwater converges as it runs down. Large valleys unload lots of water, and the splash and splatter may be destructive to a porch or stoop roof below. Snow and ice fall off roofs in dramatic fashion, too, and it pays to imagine where the stuff might land.

A steep-pitched roof that bottoms out on a shallow-pitched roof is another invitation to problems in snow country. Snow and ice build up at the bottom of the steep roof, and the draining of this pile is inhibited by the shallow roof. The joint between steep and shallow must be carefully flashed with a hidden leak shield, which I'll explain later on. With care, it's possible to build a leak-free roof of almost any type.

Ventilation

The design and specifications for a roof should embrace the whole picture: not only the shape but the shingling material, ventilating schemes, configuration of trim and overhangs, and the use of the space under it. Architects' plans show this information in an "X-ray" side view called a section view. This chap-

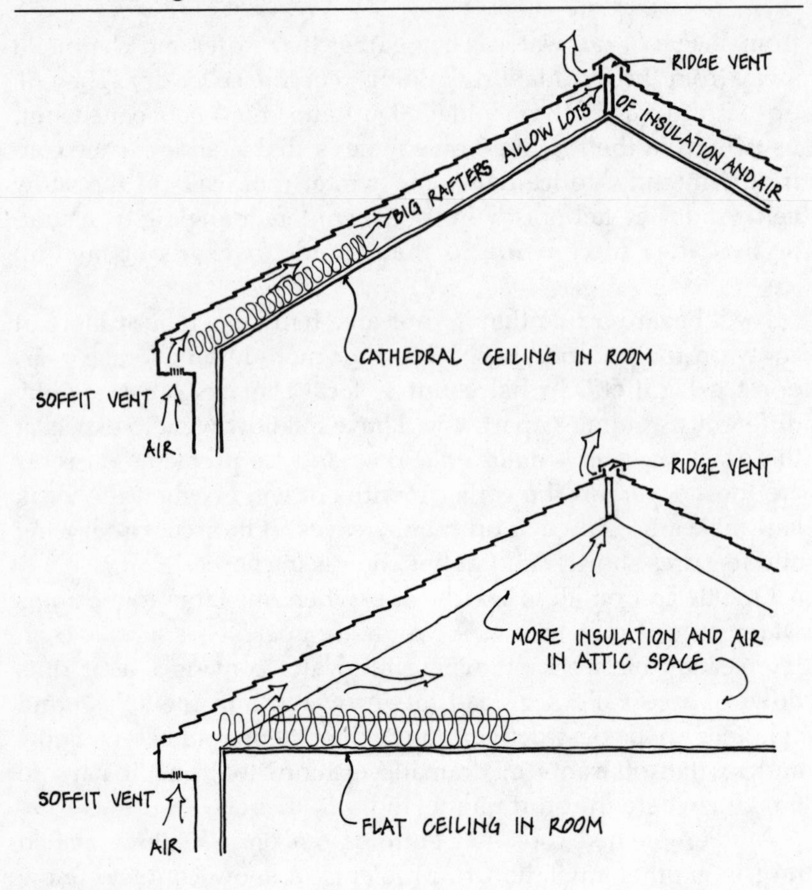

CATHEDRAL AND ATTIC VENTILATION

ter may read like a foreign language to you, but you should learn something about the subject. It will help if you refer to the illustration nearby.

First off, is the roof over an attic or a sloped (cathedral) ceiling? An unoccupied or storage attic should be designed to stay roughly the same temperature as the outdoor air. The ceiling below, not the roof, is insulated. In a cathedral-ceilinged roof, the insulation goes between the rafters, so the rafters are deeper to hold lots of it. There must be enough ventilation between the insulation and the roof sheathing to keep the roof close to the outside temperature.

You need a lot of ventilation. Most rules of thumb for
tion suggest a free-ventilating opening equal to 1/300 of th
floor area. This is a bare minimum, and not enough wh
gable louvers are specified. Screens cut down the free air by a third, and more when they are dirty, a common condition. I like to see around 1/100 of the floor area in free air.

If you plan to use a whole-house fan for cooling, you'll need larger ventilation openings yet. These fans mount in the attic floor and pull air up through the house from the presumably cooler cellar or ground-floor windows. They work well if the air can escape the attic fast enough through the ventilation openings. We usually make a ceiling-mounted insulated cover for the fans; in winter, they are great leakers of expensive and moist warm air. Specify a slow-speed fan, which is more expensive than the standard issue. The slow fan is much quieter, so you'll use it more, and lasts much longer besides. Though these fans can cool the house, the attic needs its own air movement. I harp on roof ventilation to all my customers. There are three reasons I do.

Reason one is comfort. The summer sun bakes the shingles on your roof. Much of their heat is radiated down from the roof deck into the attic, or into the rafter space in the case of cathedral ceilings. Inevitably, some of that heat passes through the ceiling into the living spaces. Even if it didn't, the fact that the attic is hotter than the rooms below keeps those rooms hot. In a cathedral-ceilinged room this problem is especially vicious.

Ventilation keeps outside air, almost always cooler than sun-heated roofing, moving between you and the shingles. In a roof designed with air intakes in the eaves and air exhausts in the ridge, the cooler air will circulate even on windless days. The buffering effect of this air moving between the hot roof structure and you in the room below is worth a ton of air conditioner capacity.

Reason two is economy. That reduced air conditioner demand is certainly worth dollars in your installation and maintenance budgets. Shingles on a ventilated roof last years longer than those remorselessly broiled without benefit of underside cooling. Condensation in an unventilated structure under the roof causes expensive rotting and allied insect destruction. Condensation also waterlogs your attic or ceiling insulation, sometimes rendering it all but useless. Condensate can even drip into the rooms of your house, acting for all the world like a roof leak. Naturally, repairs are costly as well as inconvenient. Condensation problems are not rare in modern tight houses, but a real factor to be designed out.

Reason three, the one builders fear most, is water damage from ice dams. Ice dams start when snow sits on the roof. Heat escaping from the building melts the snow from underneath, and the snow insulates the meltwater as it runs down the roof. When the water gets to the eaves, or any unheated part of the roof, it freezes. Some drips off as icicles; the rest builds up little by little on the edge of the roof. As the ice mounds higher, it dams the new meltwater to form a pond just inside the roof's edge. The shingles can shed only *running* water, not contain a pond, so they leak. The water makes its way into the structure, first the overhanging eaves, then inside the house. Shingles at the eaves are often damaged by desperate homeowners armed with axes or shovels attempting to break up the dam to let the pond out. All in all, ice dams are a big pain, the more so for forming in weather in which it's difficult to fix them.

Several after-the-fact remedies can relieve the symptoms of an ice-forming roof. You've probably seen snow belts, aluminum panels nailed onto roofs in place of the bottom few rows of shingles. The panels are crimped together, and form an impermeable barrier to ponded water. Further, the slippery aluminum helps ice fall off soon after it forms, generally with good result. The panels must go far enough up the roof so they are substantially over a heated section of the house, so that ice dams don't form above the panels. Snow belting looks lousy, but it does seem to work.

Another ex post facto fix is electric resistance cables zigzagged along the eaves. These cables plug into house current and warm up to fifty or sixty degrees to melt the ice immediately around them. The zigzags should be attached so they extend up through the likely dam sites and onto roof that is over heated areas. This way, meltwater can run down the channels the cables form and off the roof. Of course, the cables take energy to run, and must be plugged in and unplugged periodically. If your house has

gutters, they too must be cabled so dams don't form a
them. In extremely cold climates, downspouts are heat
make channels through them. Most people dislike the loc
the cables, and who wants one more maintenance worry? So what do you do? As usual, the best cure is a good design.

The first line of defense is keeping the heat in the building. The more insulation there is between hot and cold, the less gets through to melt the snow on the roof. In theory, this step should be enough. In practice, some heat still escapes, and the sun's melting and winter rains conspire to keep some water running down the roof anyway.

The next step, more insurance than anything else, is to build the bottom few feet of the roof watertight. When water ponds on a roof, it finds its way through the shingles and runs along the roofing nails into the structure. The object of waterproofing is to catch the water seeping in this way. Caulking around the nails after they are driven in is futile; the leaks are out of sight. The cure is a membrane under the shingles that seals around the nails as they are driven, leaving an unbroken barrier to turn the floods.

The membrane can be anything that will remain stable over the years and maintain the seal around the nails. Proprietary products, such as W. R. Grace's Bituthene, are generally made of neoprene, a synthetic rubber. The homemade version Apple Corps uses, a sandwich of felt, roof cement, and felt, also works well. The membrane has to extend far enough up the roof to seal any potential leak. The shallower the roof's pitch, the wider the pond behind the dam, so more area must be sealed. On houses without broad overhangs, a three-foot-wide band of the membrane is enough, and most products come that wide.

Skylights present their own problems, though the theories are the same. Skylights lose more heat than surrounding roof areas and produce more meltwater. At cathedral ceilings, special pains must be taken to vent air around the skylights, as they block the normal air flow route up the rafter bays. I drill big holes in the rafters immediately around the skylights in a cathedral to guarantee air flow. A wide skylight should have membrane waterproofing applied to the roof area above it.

I also use the membrane in valleys, where the slopes of two

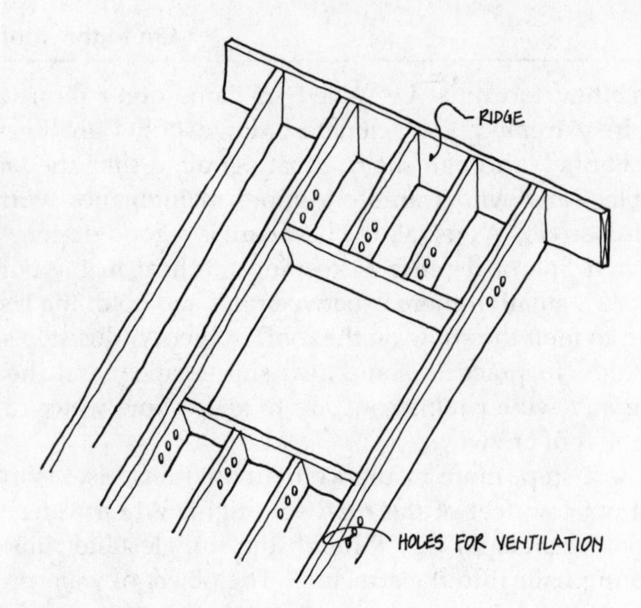

SKYLIGHT FRAMING

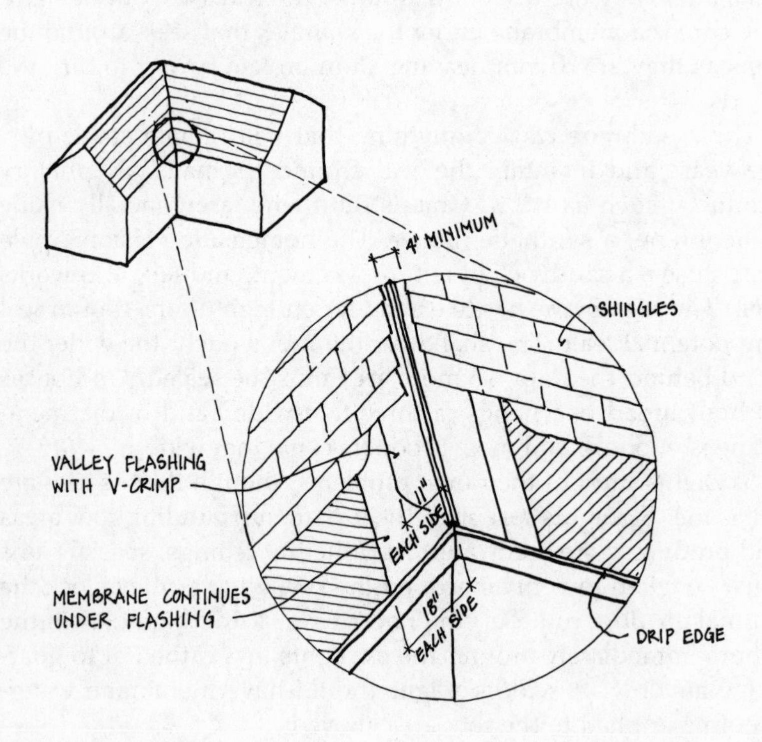

VALLEY

roofs intersect. They are a great place for snow to collect and a miserable place to remove ice from. So I run the membrane up the valleys at least one third of their length — all the way up on low-pitched roofs.

The last and best defense against ice dams is ventilation. The snow on the roof won't melt if the roof's below freezing. I like to see a two-inch-wide continuous screened slot at the eaves letting outdoor air into the roof. Exhausting that air at the highest point of the roof, the ridge, makes for the best air flow. Failing that, the best exhaust is grilles set high in the end walls. It takes huge louvers there to create enough flow, about 1/200 of the attic's floor area at each end. For the designer, the ridge vent is often easiest to work with.

Various ridge vents are available that let the air out while providing rain protection. Aluminum ones look lousy but work well. Some kinds are made to be shingled, a more expensive but prettier detail. I've made some myself in various shapes and sizes. Somehow I always end up using the metal, and cursing it as I nail it down.

The roof's surface

We usually use a metal drip edge on the roof's eaves, and this goes on first. The drip edge reinforces the edge of the shingles where the water drips off, the most vulnerable point on the roof. Old-timers used to use a double run of red cedar shingles as a starter course at the eaves, and some builders still do. As it ages, the wood is susceptible to rot and fungus, to deterioration from freeze-thaw cycles, and to damage from ladders. It gets soaked and wicks water up into the underlying roof structure and trim. It makes a good-looking drip edge, but this is an area where technology has wrought an improvement.

I prefer to use a metal valley, too. The alternative, a woven valley, is one sealed by the weaving together of shingles from the adjacent roof pitches. It's hard to make this valley look neat, and debris tends to collect in it easily. It means the roof can be completed by shinglers, who need little training in metalworking, so it's cheaper and faster.

A metal valley can be exposed or hidden. In the exposed version, a tapering band of metal appears down the roof intersection. The same metal can be covered by the shingles, resulting in a cleaner roof line with no metal showing. I prefer the exposed one because it cleans itself of leaves and other debris. As always, there are good valleys and bad valleys. I'll follow a good one as it starts up from the bottom.

Since the drip edge is galvanized steel and the valley is another metal, we separate them with felt (tar paper) to prevent galvanic corrosion. The membrane I talked about runs up the valley a few feet, or all the way to the top if it's a north side or the house is in a severe climate. Since the membrane is three feet wide, it extends well beyond the metal valley, which goes on next.

First choice of metals for the valley and for most other roof flashings is lead-coated copper, a sheet of lightly tempered copper surfaced with lead. Plain copper is nice, and its shine announces your good taste (in picking copper over the cheaper aluminum) to the world. In some places copper gets discolored, though, and can stain anything it splashes onto. Lead is more resistant to the elements and, with a backing of the tougher copper, has a long life expectancy. And, really, if it's class you're after, your subtlety in disguising your copper with a dull lead coating beats your neighbor's sparkling arrogance hands down.

Other common metals are aluminum, plain copper, and galvanized steel. Aluminum starts out shiny but dulls in a year or two. It's fairly cheap and relatively easy to work, even in the .028-inch thickness that I consider the minimum. It gets brittle with age, and should be replaced each time the roof is. Copper lasts through several roofs, as well it should, costing several times more. Galvanized steel is rarely used these days. So, use lead-coated copper if you can swing it, pick copper as your runner-up, and aluminum as third choice. The shingles, neatly cut and not nailed through the metal, complete the valley.

Where the side of a roof's slope runs up against a vertical wall or a chimney, you need step flashing, which refers to the orderly march of flashing pieces up the roof, one to each course of shingles. Use the same metal you specified in the valleys.

You have a choice of how to treat the ends of your gable roof. Some designers like to see the metal drip edge run all the way around the roof, making a continuous shadow line at this plane.

On asphalt roofs I use an asphalt shingle turned upside down so its granular surface shows when you look up. This is neat, easy to do, and not expensive. The metal edge always seems to be bent or kinked in an unsightly way, and can shelter small wasp congregations. Besides, it must be painted to blend into the trim, and the paint won't always stick.

Most houses are roofed with fiberglass shingles. They are made, in pieces three feet by one foot, from a fiberglass mat coated with asphalt, with colored mineral granules spread on top. Hard to obtain now are asphalt shingles made with a rag mat. I like them better, especially for winter work, but fiberglass is taking over. Either will keep your house dry, and, confusingly, both are called asphalt shingles. They come in a staggering variety of colors, shapes, thicknesses, and weights. They have their problems as roofing, but everyone knows them, and they're relatively cheap. They last from fifteen to thirty years, depending on their weight and where they're installed. Shingles are laid over felt, which is omitted from cut-rate roofs but shouldn't be from yours. In chapter 13 I will discuss requirements for roofing porches and other low-pitched structures.

I usually recommend tabless shingles to people choosing asphalt shingles. Cutouts (narrow slots) divide one long shingle so it appears to be two or three pieces, or tabs. The pattern of many lines enlivens the look of the roof. The cutouts localize wear by rain and sun and mean that tab shingles have a shorter life than their tabless brothers. I like the look of tab shingles, but recommend those without.

Shingles come in several weights, the difference being the thickness of the granular surface and the weight of the asphalt-impregnated mat. Standard roof shingles weigh around 220 pounds, and heavyweights 290 to 330 pounds per square. (Remember the term "square," because all roofing products are discussed that way. A square is a ten-foot-by-ten-foot area of the roof plane.) Plain, heavyweight, tabless shingles are a good bet on a new house you plan to keep because they last much longer than the standards.

Related to heavyweights are architectural shingles, favored by their namesakes because they suggest the appearance of wood or tile roofing, though they are made of fiberglass and asphalt like any ordinary shingle. These hybrids cost more than heavy-weights. Most architectural shingles have a life span no longer than that of a standard lightweight shingle, but they do dress up a roof. You pays your money and you . . .

Probably 95 percent of new houses have fiberglass-asphalt shingles. There are many other materials for roofs, from thatch to concrete tiles, from galvanized-steel sheets to wood shakes. Designers, builders, and homeowners are forever trying these other ones, striving for the ideal combination of permanence, good looks, and economy. There are benefits and problems associated with each, and I will run through a quick summary.

Metal roofing started out on barns, which have long, unobstructed stretches of roof and beg for large pieces of roofing. Steel goes on fast and lasts a long time if it is painted regularly. It does move around a good deal in response to temperature variations, stretching the holes around the nails. Nails sometimes work their way out of drying rafters, loosening the steel that must be held down to be watertight. A barn roof that leaks a little every so often is par for the course, but a leaky house roof is a catastrophe.

Most of the same attributes apply to aluminum roofing. Aluminum is lighter and more fragile, easier to handle but more likely to crack or deform under impact or pressure. It has the same nailing problems as steel. You can get either in factory-applied colors, the only way to go if you want a painted roof. I prefer steel because it stays put better and lasts a long time when painted. Neither is good when the roof contains valleys, vent pipes, or dormers. Both steel and aluminum work best on the simple vastness of a barn roof.

More exotic is terne metal, an alloy of tin and lead applied to a stronger base metal, like steel. Its dull finish endears it to some architects and is the origin of its name, from the Middle French for "tarnished." Terne sheets have a single seam at each edge and a flat center section. They are tough to bend, like steel, but can be soldered for tricky flashings. You need a specialist to apply the stuff, it costs a lot, has a quite distinctive look, and lasts a hundred years. Terrific.

A relatively recent entry into the roofing market is corrugated asphalt sheets, an example being Onduline. This comes in a couple of sizes, large squares, and can be installed by someone working alone. Onduline comes with a thirty-year guarantee, though I have seen eight-year-old installations with erosion of the surface layer. Onduline is tough to lay neatly, as it is easily distorted in storage. It is quieter in sun and hail than a metal roof, and since it has regular corrugations rather than the peaks and plains of metal, many like its looks. I don't recommend Onduline for houses, for the same reasons that apply to other sheet roofing products, but it makes a handsome utility roof on a barn or shed.

Early American settlers' roofs were of bark or shakes. Shakes are short boards split from sections of logs and crudely smoothed. Rows of them were fastened to poles (called purlins) running crosswise of the rafters, each piece in the row covering the joint between two in the row below. This was not a perfect roof, particularly when a wind pushed rain and snow through the many interstices. The more trouble the roofer (often the homeowner-to-be) took smoothing the shakes, the less water found its way in; craftsmanship had its direct reward.

The rough shake could be made with hand tools, a saw and a froe and mallet. Then and now, shakes are split, wood shingles are sawed. When sawmills overspread the nineteenth century, accurately cut wood shingles, the elusive goal of most shake splitters, could be fashioned quickly. Hundreds of bolting mills (bolts are shingle-length logs) produced trillions of shingles, the roofing of the hoi polloi. Wood shingles were thinner than shakes, so more could be made from a given tree. And the more uniform shapes made for a tighter roof.

Wood shakes are used nowadays on fancy custom houses from California to Maine. They cost plenty and are tricky and expensive to install. They aren't so good in the shady woods, where funguses attack them. They're prone to burn in dry weather from fireplace sparks. But they provide an unmatched look combining ruggedness and purpose and nostalgic charm. I don't recommend them much in my snowy climate, but I've put some on for those who can accept no substitute.

Wood shingles, sawed to regular thickness and taper, are much more common. They suffer more from funguses than do shakes, being thinner and more tightly laid. Before installation, either can be dipped or brushed with a fungicide, but this is a stopgap and can't last many seasons. A good shingle roof can last forty years in an open location with mild winters, but a heavyweight asphalt roof can come near that life span and costs much less. And, under unfavorable conditions, asphalt outlasts the wood three to one. Face it, you buy a wood roof for its looks and romantic qualities, not to be practical. No asphalt imitator looks the same.

There is a debate among roofers about the best base for wood shingles. Some favor a standard plywood roof deck covered with fifteen-pound felt. The felt is a backup for the shingles, and can catch errant trickles and send them on their way. The option is skip sheathing, boards spaced a few inches apart. Adherents of skip sheathing maintain that their system allows the shingles to breathe from both sides, so they absorb and release rainwater without destructive curling. I can see little practical difference; the aerobic shingles may last a bit longer, but when they leak they leak. The shingles on felt squirm around more in their fettered atmosphere, but keep the rain out of the house about as long. I instinctively favor skip sheathing, but like the bracing plywood gives the roof plane. Talk this one over with your builder during the pre-bid talks. Either method, properly executed, will give you a good wood shingle roof.

After the Civil War, New England's merchants graduated from wood to slate shingles; properly installed as a house's original roof, they could last centuries. The later years of the 1800s brought itinerant slaters to New England. These fellows, the smooth-talking tin men of their day, worked from the nearest railway siding. Luckily, they mostly had a good product to push, and much of their work endures today. Unfortunately, many slate roofs were nailed over old wood shingles, denying the slates the firm base they deserved and required for an enduring installation. Slates are great, but tough to work on without care and training, and dangerous to walk on when underlying wood shingles shift and powder over time.

A new slate roof is a costly item, with materials at five hundred dollars a square and skilled labor rare and pricey. I've put up some small roofs with used slates and think they're great. Tile roofs come under the same category, though the price is lower.

Clay tiles are fragile and expensive, and concrete tiles give a similar effect in a tougher product. Either must be properly installed, and neither is recommended in areas of freeze-thaw cycles. Most people end up with asphalt shingles, a few with wood shingles and shakes, and a scattered elite with slate, tiles, and other exotica.

One reason for the popularity of asphalt shingles is they are forgiving. Except at roof junctions and penetrations (chimneys, vent pipes and so on), they can be laid quick and dirty and still turn the rain, so a builder can do the tricky flashings and valleys and let his apprentices lay the bulk of the shingles. It takes patience and thought to make a neat-looking job, but many roofs are up out of sight anyway.

Roofing is quintessentially a summer job, and often gets inexperienced "summer help." If you're pushing your builder for a cheap job, don't expect much fussing. Roofs with broad expanses of repeating patterns show up slight deviations dramatically. If your roof is in plain sight and is a major feature of your house's design, request experienced and careful roofers, preferably your builder's own crew.

When your roof is shingled, it's time to get ready for the door and window deliveries. Builders wait until now so the ebullience of the roofers doesn't net you a broken window or scratched door. The pace of the job is still full tilt, and fifteen thousand dollars' worth of windows can go in in a day or two. A new phase of the job is coming soon, the interior subcontracts. For now, let's get windows and doors installed.

9

Windows and Doors

YOU'LL PROBABLY TALK about windows and doors as much as anything else on your plans. These mechanical holes in your house figure prominently in its budget and your life inside. The type, placement, and size of the windows and doors you specify affect light, ventilation, traffic flow, heat loss, views, and status. You read in chapter 7 about where they go. You can read a lot about the devices themselves here, though your architect might exert more influence on your choices.

Traditional double-hung (sliding up and down) windows, inexpensive and durable, persist as old standbys. Most new double-hungs are somewhat stiff when new, and tough to use at arm's length — say, over the kitchen sink. When closed, they are slightly more drafty than other kinds. Fully open, they still admit only half the breeze of an equal-size casement window. The top sash (the sash is the separate, moveable frame holding the glass) of some double-hungs opens only partway, or not at all when half screens are in place. Some of these screens are awkward to remove from the inside of the house, a possible problem on second-floor windows. On many styles the sashes tilt into the room for washing, though this is not an operation for the faint of heart; it requires pluck and strength to yank the devils out of their tracks.

On the other hand, double-hungs are good at some things. Since the sashes slide in place, nothing protrudes into the room or outside to spear unwary passers-by. You can crack the window open slightly on a humid morning, and afternoon showers won't soak the floor or penetrate the sash itself. Mechanical parts are few and slow to wear out. Everybody understands them. Mainly, they embody the traditional appearance certain house designs call for; they're what many of us mean when we say "window."

A variant of the double-hung is the sliding window, a hung window turned on its side. All the drawbacks of double-hung windows are present: draftiness, difficult operation at arm's length, only half open when open. The saving graces of traditional looks and a little ventilation during rain showers are missing, though. This is a mongrel. Its only visible virtue is its low price.

Casement windows are pretty good. They close tight, like a new door, and open easily. Turn a crank, and the whole opening lets in the spring air. In the right place, the open sash even diverts into the house breezes that might otherwise slip by. On most units, both sides of the glass can be cleaned from inside without the tricky derailing of the double-hung sash.

Casements won't work everywhere. When a porch or deck is outside the window, an open casement is in the way and dangerous. A casement window open even a couple of inches in a rainstorm can funnel damaging water into the house and into its own working parts. Casement operating mechanisms are rugged but more complex than those of other windows. In some spots you'll be able to reach the crank but not the sash latch, an annoying difficulty. Finally, casements just don't look like hung windows, and are hard to integrate into a traditional design.

A casement turned on its side is an awning window. Awnings obstruct a deck just like a casement, but can be left partly open in the rain without damaging anything. They are easy to operate and weathertight. Their long dimension is usually horizontal, much as the casements' is vertical, so the style works only with certain designs. And, like the casement, they are more expensive than hung or sliding windows.

If you buy either casement or awning windows, stay away

from the largest sizes in the manufacturer's range. The mechanisms are often designed for the normal-size sash, and some can barely manage the biggest ones. In the small sizes, you are buying more frame and mechanism than glass compared to the larger units. The best value and operation comes with the middle sizes.

Once you and your architect have picked the type and size of windows, and placed them in the plans, you have other choices to make. Not all manufacturers offer all the options, and the availability of one or the other may push you to a certain brand. The following list should help you decide what to order.

Glass is the major feature of the window, and choices here give you much to think about. Glazing means the mounting of the glass in the frames or sashes, and is also used to describe the glass itself. Single glazing, one pane of glass in the opening, is suitable only for an unheated building or room, or for a partition or window between two equally heated rooms. Since a single sheet of glass conducts heat readily, it won't do between you and the outdoors. Household moisture condenses on the cool glass and trickles down into damaging puddles on window sills and sashes.

The next step up is double glazing, now standard in most new houses, at least in colder regions. There are two ways of getting double glazing, and the goal of both is to trap air between two sheets of glass. (Air, held immobile, is the operating agent of any insulation.) The first double glazing was insulated glass, two sheets of glass fused or joined at their perimeters with the space between (about one-quarter inch) a vacuum. This worked great until the joint leaked or the glass cracked a little. Then temperature changes during a day or over a season would suck room air into the space, and the moisture it carried would fog up the windows. That condition is called a seal failure, and is the scourge of thermal or insulated glass. It's not curable, and means replacing the glass, expensive when it happens all over the house.

To get around such failures, manufacturers are no longer leaving a vacuum between panes of glass. Now they just clean the glass, put moisture-absorbing crystals in the metal separator strips, and hope for the best. It works. Other window makers specify a separate piece of glass, called a double-glazing panel or storm panel, which can be removed for cleaning. Clips hold the panel onto the sash to form the dead-air space. This works too, and never has a seal failure. The window company can make single-glazed windows for those who want them, and clip panels on for those who need more insulation.

Another virtue of removable panels is apparent only *in extremis.* If a well-placed iceball or poorly placed partridge shatters your insulated glass window, you're out of luck (and heat) until you can get it fixed. With panels, just clip one from a similar-size window into the broken sash; you're almost good as new while awaiting repairs.

Some of the clips are flimsy, though, and seem to be strong enough for only rare use. A thin sheet-metal or plastic clip won't take too many off-and-on cycles before it doesn't hold the glass securely anymore. The Pella system of clips going through the panel into slots in the sash seems to be the strongest. Also, if the panels are on the inside of the window, the clips as well as the gasket are out of the weather, and the panels are certainly easier to remove. One good thing about exterior panels shows up on Marvin windows. On their "authentic divided lite" models with fragile wood strips (muntins) separating small panes of glass, the exterior panel keeps the weather away from the muntins.

Triple glazing is also an option in some brands. It combines an insulated glass sash with a single-glazed storm panel. Triple glazing is effective, though since each pane of glass stops 10 percent of the light going through it, you can't see through triple-glazed windows as well. Also, the triple-glazed sash is heavy, especially in the larger sizes.

I prefer the latest development in window glass, low-emissivity, or low-E, glass. In this method, a coating that reduces the radiative emission of heat from glass surfaces is fused onto or into the glass. Low-E windows have similar or better R-values than triple-glazed windows, without all the bulk or reduced light transmission. Here is new technology that really works. If the low-E coating is on the panel glass, so much the better, as repairs will be cheaper and easier than those to the primary pane.

Building codes in some states require safety glass in windows next to doors and hallways. Safety glass isn't the wired kind you remember from school, but glass tempered to break into bits rather than spears. Safety glass may not be available in low-E, but the code comes first. Plexiglas, the plastic answer to jagged glass shards, doesn't stand up to human wear and tear, much less Rover's toenails. Get safety glass — don't mess around.

Some windows come without screens; you have to order them separately. Casement and awning screens are only one style, but double-hung screens may be available in half or full size. Half size means you can't open the top sash; the screen sits in the track below it. Full screens cost more, allow you to open top and bottom, and haze the view through them both. One isn't better than the other. Some companies make their screen frames in different colors, so you can complement your trim color with the screens.

Windows normally come with their outside trim attached, and can be nailed right in place as soon as they arrive on the site. Some companies — Marvin is one — will customize your order with various sizes of trim. If your plans call for nonstandard trim, and the maker won't supply it, the builder can install his own, though it's somewhat more pricey. Anyway, if your manufacturer of choice does offer the trim style you want, be sure to order your windows that way. (Use the summary at the end of this section as a checklist for completing your window order.)

When you order factory-applied standard trim, some manufacturers offer various finishes on the exterior parts. Exteriors can be bare wood, primed wood, primed and painted wood, wood with vinyl cladding (covering), wood with metal cladding (usually aluminum), solid vinyl, solid aluminum, and other permutations. I prefer wood or wood clad with metal. Vinyl is brittle when cold, flexible when hot, and can't be altered to fit different house styles. Solid vinyl or aluminum windows are often heat-robbing, fixed-style, cost-competitive units that are hard to justify in a well-built house. All-metal windows, even with thermal breaks to reduce heat loss, just aren't the tight, naturally well insulated units wood windows are.

When I order wood windows, with or without exterior trim, I get either primed or painted units. The factory finishing or priming is well done and smooth, and cheaper than having a painter do it. If your paint scheme calls for a finish paint not offered on the windows, or you plan to install your own outside trim on them, getting the primed units is best. If the factory paint comes in your color, order it.

Even if the exterior of your house is to be stained, *you should still paint the windows.* Stain isn't enough of a finish for windows. It leaves no film to turn water away, and can deteriorate to the danger point, leaving crucial window parts vulnerable without showing much sign of its failure. Windows are just too valuable to trust their long life to a product that can so inconspicuously lose its effectiveness. You need the film the paint provides, both to turn water and to show it needs maintenance. Don't economize on this one.

Fixed, nonopening, windows are made by all of the good-quality manufacturers to match or complement their standard windows. By all means order the matching fixed units if they are available, rather than have them built on the site. The style, finish, and quality of the manufactured units are the same as the regular windows, and they don't look half-baked. Also, the product is guaranteed. The glass in most fixed windows is insulated. With site-built fixed-glass windows, you run the risk of the glass supplier's voiding his guarantee based on your (supposedly) improper installation.

Specialized manufacturers can make windows for you in almost any size and shape you can dream up. Colonial reproductions can get spidery multipaned twelve-over-eighteens. Bold contemporaries can feature curved units filling the corners of the spa room. Most unusual types and sizes are quite expensive, though, so cost-conscious designers tend to use novel groupings or placements of standard windows to achieve drama. Any uncommon installation will likely be at least as troublesome as it is costly. Order exotics if you wish, but do so with your eyes open.

It's best not to overlook proper installation. To work satisfactorily and last long, a window should be adequately supported and securely fastened, sit in a flat plane, be plumb (absolutely vertical) and level, have water diverted around it, be treated gently when installed and used, and be finished well. Most companies' windows can be mulled — ganged up in a larger, composite unit. If your plans call for mulled units, use the manufacturer's version, to keep from voiding the warranty.

The reason for this care is to give the window the best chance of operating as it was intended to. If the opening is not flat, double-hungs may hang up and sliders may not slide as they should. A twisted opening also forces open joints in the window casings and frames, promising a short service life. We've installed many windows in old, crooked houses, and it takes a lot of fussing to make the windows operate properly. Most troubles can be traced to bad framing.

Water on the outside of the building should stay there. It shouldn't appear inside where you can see it, or in the structure where you may not see it until the carpenter ants move out. Windows must be protected from water by flashing, which diverts water running down the side of the house, sending it around and out past the window.

Early American settlers used wood flashing over windows and doors. It was what they had, and it worked. I've installed it myself, though often with a hidden metal back-up. A wood flashing, kept well painted, should last as long as the siding of the house. Wood flashings are mainly used with horizontal sidings, like clapboards. They're a bit fussy to install, and aluminum is cheaper, so they aren't often used anymore. They do look substantial, though, so in some designs nothing else will do.

On some colonial or Victorian reproductions or styles, exterior window trim is quite bold or fancy. Drawings may call for tiny roofs over each window, or layers of moldings to accent the openings. It may be flight of fancy or a utilitarian scupper, but the protection these built-out top casings give the windows is great. We must be sure the moldings themselves are guarded against water leaking in. A metal cap flashing the whole works is a requirement. Be sure the doodads don't interfere with installing or removing screens or storms.

To sum up, figure out your window choices, starting from the top. Type comes first; choose hung, casement, sliding, or awning. Under style, consider outside trim, muntins (strips dividing up the sash), locks, and hardware. Glazing choices are single, double, storm panels, triple, low-E, safety, tinted, and so on. Screens are half or full, of a certain color. Most often, by the time you have selected your way through this list, the manufacturer will be obvious, chosen by default.

Your order information should include the count of each size unit, jamb extensions if necessary to fit your wall thickness, exterior trim or color, screen type, left- or right-hand opening for casements, muntins and/or mullions (the joints between combined units — the product of mulling) where necessary, and any factory finishes you want applied. Plan ahead; you may have to wait many weeks for specialized units. Your local dealer may not have what you need in stock, and a three- to six-week delay for ordered windows is normal. Make sure they're not delivered until you're ready for them, as they're too expensive and fragile to be used as sawhorses at the job site. Windows cost a lot.

Doors and windows are usually lumped in a single category by builders. They often are delivered at the same time, just as the roofing is completed. They must be installed before the siding, and the builder is anxious to get the house closed in and be able to lock it. Like windows, doors are expensive and vulnerable, twisting and sticking and leaking all out of proportion to their price. Unlike windows, doors are opened and shut constantly in all seasons, scratched by pets, kicked by grocery carriers and irate teens, and expected to look welcoming, operate smoothly, and require no maintenance whatsoever.

Keep in mind my recommendations from chapter 7. Shelter your doors from the weather. Make one door, at least, visible to your guests. One door should be handy for your most frequent trips in and out. Don't put a slider anywhere you want to travel regularly. Step down when you step out. Many door problems are actually products of the design.

Every installation principle that applies to windows goes for doors. Flashing is especially important unless the door is under a wide overhang. Doors need flashing underneath, too, since there's often lots of water around them. Your section drawing should show metal that is bent under the threshold, then out and down. Two caulk lines, at inside and outside edges, should be evident. The door's threshold should sit flat on the plywood decking. The door should be shimmed into alignment all around, not just nailed into the rough opening through the casings. Though a good builder will do all this stuff, there's no harm in having it show on the drawings.

A common problem when installing an outside door is setting it the correct height above the floor. Doors are put in when only the rough floor is in place. The thickness of the finish floor must be accounted for, plus any mats or rugs you plan to have inside. Since the door's threshold normally sets the door only about 1¼ inches off the rough floor, a wood floor and mat, a carpet and mat, or even just a thick carpet sometimes will end up too high for the door to clear. You can check the clearance by scrutinizing a section drawing of the door area, and correct one that's too low by gently questioning the designer.

Sliders and French and patio doors require even more scrupulous installation. I've seen many sliders nailed only through their outside flange or trim, not solidly through their frame. They just can't last, especially with any rough treatment. French doors are the toughest because so many places must be fitted and all work together. Even a carefully framed new house has some discrepancies in its framing; it's the independent nature of wood. I usually figure on half a day to fully install a steel single door, three quarters for a patio door, and a day and a half for a French door pair, with an extra day for wood storms. A professional door subcontractor can put two exterior and fifteen interior doors in each of two houses in one day! "Shims" and "fitting" are foreign words to him. You have to pay for the door itself no matter who installs it. Only a good installation makes an expensive door worth it.

10

The Subs Arrive

WHEN THE DOORS and windows are all installed, the job shifts gears. The outside surfaces of the house are far from finished, but it's time to get started inside with the subcontractors, "the trades." While carpenters take on the trim and siding, the foreman organizes what's called roughing-in. This means installing the chimney and all the electrical and mechanical parts that will be covered by the interior finished surfaces.

Ideally, each sub has the house to himself during his installation, and roughing-in proceeds in sequence. The mason arrives first, so nobody will use the chimney space for ducts or pipes. Next, HVAC — heating, ventilating, and air conditioning — is the first mechanical sub, because ducts are the biggest items to be threaded through the framing and the hardest to adapt to other subs' installations. Then comes plumbing, with its rigid pipes and strict code rules. Wiring is pretty flexible, and can be run around pipes and ducts where necessary. Alarm systems, intercoms, and telephones use even thinner wires and go in last. In a perfect world of no time pressure, this approach yields the best results.

Normally, though, tradesmen have their own schedules to keep and other jobs to work on besides yours, so the chance of programming all these workers in ideal order is slim. It's not unusual to have ten tradesmen, including assistants, to rough-in a house. Even if you could schedule them all to come at separate times, there would still be one place where Frank must box-in a duct and Marty must run a vent stack. Someone must mediate. May I suggest a good general contractor?

The most hectic time for the general or his foreman is now. Every tradesman needs a share of his attention. The architect and owner will visit frequently to assure themselves of getting what they asked for. Drawings of duct locations, waste-pipe runs, and electrical outlets are sometimes at odds with the logic or experience of the installer. A capable foreman can see that the architect's intent is followed while giving each tradesman enough freedom to install a system he is satisfied with. A foreman must be tough but cooperative, expect high standards, yet be objective about the goals and tone of the job.

If you are an owner-contractor, this is your worst time, too. It's often difficult to separate your anxiety about the project from your treatment of the tradesmen. With good intentions informed by too little experience, you can alienate a tradesman to the point of spoiling his installation. You may be unaware of important peripheral issues that aren't related to your job but matter to the subs. Did you hire a plumbing and heating contractor for the HVAC and a competing plumber for the plumbing? Was there trouble the last time these two worked around each other? As a novice you may never learn directly of these difficulties. Overcoming them is part of what you didn't pay a general contractor to do. He or his foreman is your escort and advocate in this thicket of decisions, personalities, and conventions.

Even if you're not acting as your own general contractor, you should be available to visit the site often during roughing-in. Final placement of many fixtures and devices (outlets and switches) is determined now. As the walls and openings are framed, the shapes of the spaces should become real for you. I like to walk through the whole house with the owner and each tradesman, discussing the work slated for each room. I find this informative for my work to come, and I value the overview and detailed awareness I get of the mechanical layout.

The walk-through is a convenience to you, but it may result in more work for the sub, and is essentially nonproductive time in his busy day. So respect his need to move through the task quickly. Many tradesmen have their own version of what people should want, often based on what they can pick up easily from their favorite supplier. If you or your architect doesn't specify differently, that's what you'll get. A tour with each sub lets him know exactly what you're after, filling out his view of the specs. He will respect your design choices if you can tell him why and you don't vacillate for ten minutes over the location of each switch.

The implication here is that you are an outsider, at the mercy of the foreman's and tradesmen's actions and their judgments as well. You will probably feel that way during the tumult of roughing-in. The shell of your house is their workplace, their equivalent of your office or studio or shop. Most subcontractors bid their work at a fixed price, and hustle to make their living. Many have been distracted by customers at times, and may be cautious about getting too involved with you.

Back when you were still in school, you were subjected to assumptions about your home life, your parents, and your career. Sharp and dull alike, some kids were tracked toward vocations, others into college prep classes. In high school the groups split socially, cementing the division. I know from experience in both camps that the old dichotomy persists into adulthood. This may make your dealings with tradesmen tricky, and is another facet of architect-builder discord. Here *I* am assuming you were tracked away from a trade, but the fewer assumptions *you* make about those who work for you, the better you'll treat each other.

Roughing-in is a little hard to contemplate because it doesn't look like anything you're used to, yet determines the entire mechanical setup of the finished house. The rough work is more or less permanent, while the finish can generally be changed if need be. This chapter looks at only the roughing-in, and the finish installations are covered later. This brief survey does no justice to the trades' scope, but should help you to get what you want.

Masonry

Houses don't absolutely need chimneys anymore. With electric heat came the house with no mason subcontractor. Still, from a child's tilted block-with-pigtail to Victorian flights of fancy, most people must see a chimney to know a house is complete. Fireplaces are back now after a skirmish with wood stoves. Almost any fuel-burning heating system needs a chimney.

Chimneys work best inside the house, not pasted on the end somewhere. Your stove or fireplace will burn cleaner and smoke up the house less if the whole height of the chimney stays as warm as possible. Placing a chimney against the outside of the house means a chance for water leaks all the way up on both sides. Position a wood stove or fireplace out of any traffic pattern. It's great if you can arrange a short route for bringing fuel in and ashes out.

Safety is *the* concern in any wood-burning installation. Common sense tells you how far away from walls and woodwork stoves must be placed, and codes double that. You'll need a big hearth in front of the stove door or fireplace opening to catch hot embers. Your wood storage should be three feet away or more. What you don't see must be safe, too. Most codes demand a two-inch clearance between the house's structure and any part of the masonry. It's important to block that space off from the cellar in case a fire starts down there. That space outside the chimney makes a great chute for fire right up through the house. Rest assured that all this is in the code, but keep your eyes open anyway.

A smoky fireplace is a curse, and can be the result of poor design or local conditions. Wind patterns, chimney height and shape, fuel quality, and flue size all affect how smoky it is. Fireplaces with more than one opening rarely work well. It's a neat idea to be able to see the fire from different rooms, but not if it's smoking the whole time. Your architect will draw the façade of the fireplace, but the internal workings are left up to the mason.

Fireplaces heat best when there's just enough air moving through them to burn the wood and get the smoke out. The thermodynamics of air moving up a chimney flue are complex.

Lacking exact guidelines, the mason often compensates for unknowns in the equation by oversizing the flue. His goal is not efficiency but no smoke. If you are in on a dispute between your mason and your architect over the workings of your chimney, go with the mason. In the end you'll call him, not the architect, to fix the thing.

Now you get my metal chimney lecture. Masonry chimneys are costly and heavy. They take up room down to the cellar no matter where the fireplace or stove is. It's tempting to use an insulated lightweight-metal chimney. I think it's at best a temporary fix, and has no place in a nice house. Besides, it looks lousy. Though most are stainless steel, metal chimneys age quickly in the high temperatures and acid atmosphere of wood smoke. They can become severely distorted after one chimney fire, making them barely safe. If you know your chimney has become distorted, you can replace it, but it costs plenty. By all means specify a masonry chimney.

You might regard the mason and his crew as something like the excavators. They do exacting yet heavy and dirty and dangerous work. Many are given to beer and tobacco, and all their tools look used up when they're a week old. House masons may know brick and block laying, concrete floor finishing, and how to spot a good lintel stone buried in a pile of rubble. Theirs is a rough art, but art just the same.

The chimney starts in the cellar. Even with clean-out doors and furnace connections, the concrete blocks and flue tiles stack quickly up to the first floor. They form an elevated base for a small slab that is the starting platform for your fireplace. The fireplace takes some figuring and goes slower, especially when the design calls for a big or fancy masonry facing. By the time he gets to the roof and finishes the top, your mason may have been at the site a week or more. If you'll have only a plain wood stove or a stove and furnace, the chimney continues in block form right to the roof line, and will be finished in a day or two.

HVAC

Comfort is what you want from your HVAC system. Of course, you want to stay warm when it's cold out and get relief from torrid summers. You'd probably like your system to be entirely unobtrusive. The house should be under your control but able to take care of itself, too. You don't want hot or cold spots, or drafts. You want a reliable installation that doesn't need constant servicing or maintenance.

When I started in business, heating was all there was to HVAC. When it got hot in July, you opened the windows in the attic. Hot-water baseboards were the good, expensive heat; warm air was noisy but cheap to run; and electric was cheap to install. Air systems are much improved now, water systems aren't, and electric is still cheap to install. Air conditioning is gaining converts even in temperate regions.

Air delivery systems give you a chance to control the weather conditions in your house. Air can be heated and cooled, humidified or dehumidified, filtered and cleaned. A good system can be added to, so not all of these options need be bought at once. In most parts of the country an air-source heat pump system with an oil or gas back-up burner makes a reliable and economical installation. A heat pump with fuel back-up gives you a little political control over your heat, too; you can switch off the burner if the price of oil goes out of sight for a while. Heat pumps have gotten a bad name in some places because they demand but don't always get a thoughtful installation. It's pretty hard for you to judge a system by looking at it, but neat work shows up, especially in comparison. The HVAC goal is a comfortable house, and the residents are the best judges, so ask them.

Hot-water boilers are slipping in popularity even in their last bastion, New England. One reason is they don't normally come as a system. One company makes a boiler, another a circulator, and a third the baseboards. The parts sort of work together. The baseboards take up huge lengths of wall and must be cleaned regularly, even though it's awkward, given their design. It's a noisy heat, ticking and swishing with each temperature change. The whole installation is expensive, and all it will do is heat the house.

Electric baseboards work in a similar fashion, though they are quiet and cost far less to put in. Radiant electric heat in ceilings is very unobtrusive and makes a great back-up for a wood stove.

Wall-mounted electric heaters are noisy and short-lived, especially in bathrooms, where they're used because there's no room for baseboards. The price of electricity stays fairly steady, more so than that of oil or gas. Still, electric resistance heat can cost a fortune to run. I recommend it only for rooms that get occasional winter use.

A good warm-air furnace or heat pump need not be noisy. Making the delivery system of fiberglass ductboard ensures quiet. Low-velocity air is quiet, too, and the high speed and small ducts of the sixties are what gave air systems their cheap overtones. Good systems have lots of delivery and return outlets, to avoid concentrations of noise or temperature. The better installations are often designed by a careful subcontractor, not by the materials supplier who usually does the job. The inevitable midstream changes are simpler that way, and comfort, not ductwork sales, dictates the design.

The 1973 oil embargo turned lots of heads. Oil ceased to be a natural resource and became a political and financial power. We had an economic reason to be efficient. New alternative energy schemes, like hastily drilled oil wells, flared up and then burned themselves out with the failed glow of imprudent enterprise. The problematical result of that chaos is that money motivates most home heating decisions. Your own comfort and living patterns should mold your perspective more. Try warming yourself on endless January evenings with the money you saved on your heating system installation.

Plumbing

The plumbing codes are the strictest of any trade's, but each plumber and each inspector looks at his work differently. A fellow can comply with the code but leave you a third-rate system. Cheap valves will wear out, and there'll be no shutoffs to make repairs simple. Water pipes can meet code standards but be too small to deliver adequate amounts of water quickly to your second-floor tub. All a plumber is supposed to have to know is, "Hot on the left, cold on the right, water runs downhill, and wash your hands before lunch." There's a bit more to it.

Plumbers are usually responsible for designing the waste and water systems in your house. The architect provides the layout of the fixtures, but the size and arrangement of the pipes that service them is decided in the field. Plumbers wouldn't have it any other way, since they have their individual styles of work. They choose a method between doing a good job that can be easily and cheaply maintained and making a quick-and-dirty but profitable raid on your peace of mind. Your builder should get you an installing plumber who is prepared to maintain your system, too.

Within the restrictions of a whole-house contract, you can't mandate how the plumber shall be hired. Some builders, Apple Corps included, hire their plumbers on a stock-and-time basis, to make sure the job is done without one eye on the time clock. This isn't altruism but an attempt to avoid the hassles of a price-driven job. Plumbing is truly one of the basic systems of the house. It's expensive to install, hard to change, and permanent if done well. Ask your builder to hire the best plumber he knows, and to include that price in his bid. You'll pay roughly six to eight hundred dollars for each location water is available to you or your fixtures or appliances.

You may also find a plumber on your early house tours. Visual inspection can tell you lots about a plumbing system. Pipe runs should be straight, although drain pipes pitch downward slightly. All pipes should be supported often enough so they don't sag between hangers; you can tell. Copper pipe connec-

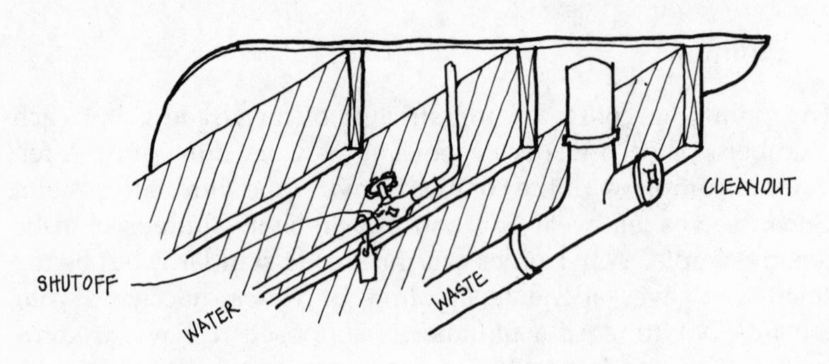

PIPES IN CELLAR

tions should be neat, without big globs of solder (the color of pewter) hanging from them. Each water pipe that runs along the cellar ceiling and then disappears up into the house should have a stop-and-waste or a plugged tee at the turn, so it can be drained. Hot-water runs in unheated areas should be insulated with pipe foam. Each DWV (drain, waste, and vent) pipe that comes down from the house into the cellar should have a tee-wye and clean-out at the turn. A very good sign is a tag on each shutoff valve stating its purpose. Neat work becomes obvious from seeing its opposite. If you see someone's work you like, ask your builder to consider hiring him.

Plumbing is one trade that technology has smiled on lately. The last couple of decades have brought plastic drain pipes, simplified cast-iron connectors, simple pipe hangers, and lead-free solder. What is considered to be quality in fixtures has fallen in inverse proportion to their abundance. The finish, strength, and engineering of toilets, especially, ain't what it was, and plastic parts are everywhere whether they're wanted or not.

Some stuff that came along wasn't as good as it promised. The big copper DWV pipes and fittings cost plenty and just don't last long. Copper water pipe works fine, though again, the thicker type M is best. Plastic water pipe has too many problems to be a force. Some new water-saving toilets just don't flush.

Plumbers get conservative with experience. They keep a pediatrician's erratic schedule, and don't like unplugging your drains in the wee hours. They learn which installations last and which don't. They pressure their suppliers with "common sense," which ultimately stifles creativity in fixture design. If you opt for new and untested systems, you buck that trend, with the risks and rewards of a pioneer. If your plumber says he won't service the gizmo, that's your risk.

The standard characterization of a plumber is of someone who is greedy but off color, dull but conniving. Just as in seeking out a urologist, you're hiring a guy to do vital work you can't fathom. You needn't empathize with his trade to know that thinking of him with prejudice won't help you get his best work. Your goal with the plumber and electrician, especially, is to establish a relationship as well as install a system.

Electricity

Electricians install everything, from the meter to the circuit breaker panel to the outlets and switches and light fixtures. Like plumbers, they must be licensed and follow a strict code, built around your safety. (Note that if you bought packaged plans, the electrical layout won't necessarily be complete or codeworthy in your locality.) Also like plumbers, they get to design their installation. You have influence in the design through your architect's electrical plan. If he doesn't plan the system, and many don't, you can do so by walking through the house with the electrician.

As you might expect, various levels of quality are possible, mostly corresponding to your convenience. For a spec house, the builder may demand a bare-bones installation. The code stipulates a proletarian setup: one switch per room, two circuits in the kitchen, the minimum spacing between outlets, and some lighting in stairwells and halls. A cheap job will also net you cut-rate switches and outlets and panels. Architects seem to favor lots of lighting in the parts of your house they care about, and much less elsewhere. Each switch or outlet means another twenty-five to thirty dollars, so you can choose. With or without an electrical plan, the electrician should have your input to install the system you want. Remember the rules for the walk-through, and know most of what you want ahead of time.

Planning is important because the rough work dictates the finish. Yet in your skeletal house you will be hard pressed to envision the rooms as you'll live in them. It just doesn't feel anything like a home inside. There will be noise, clutter, and strangers drilling holes everywhere. It's a tough time to decide whether you want a ceiling light or a switched outlet in the family room. If you know the kinds of lights you'll want and about where the switches should go, your electrician and builder can fill in the rest of the system.

For electricians, some hints of a good job are wires that are parallel and stapled regularly. The mass of wires coming into the circuit breaker panel should be separated and orderly, not a fistful of spaghetti. Every breaker in the panel should be labeled.

Three or four twenty-amp breakers marked "kitchen" are a good sign. Plenty of lights and outlets in the cellar and garage is another. If your electrician does the telephone wiring, he should apply the same care to that.

All this will cost you between a dollar fifty and four dollars per square foot of house, plus around four dollars per amp of service. A modest house with gas appliances and other than electric heat can operate adequately with one-hundred-amp service. Two hundred is the next step up, and carries bigger houses with more electrical stuff — water heaters and dryers. If you want your electricity to come into your house from underground, you'll probably have to pay extra for it, depending on your local utility. Security systems are a specialty trade, and your choices mostly involve how you will use the system.

Every sub I've ever talked to says you can tell a good installation by its neatness. Beyond that, the parameters become fuzzy, as the pro knows so many variations for any theme. Recommendations are the best source for tradesmen, phone directories the worst. Although you don't hire or pay the guys directly under a general contract to build your house, you can help get a good installation with a positive and interested attitude.

Building inspector

When the roughing-in is complete, the house gets a once-over by the municipal inspectors. Plumbing, electrical, fuel, fire detection, and sewage systems have their separate permits and administrators. This is their last chance to see that the house's innards are in compliance with the applicable codes. They have little jurisdiction over the quality of the work they inspect. Likewise the building inspector, who oversees the others and checks the house's structure for code compliance. Most inspections last only a few minutes. All these fellows will be back when the house is ready for occupancy.

It's common for an inspector to find something needing correcting. Codes are complex and change frequently. Inspectors are individuals, and more concerned about certain aspects of an installation than others. Feel free to call the inspector to get his view of the problem if you think he called for major revisions, especially in the framing. In fact, chatting with the building official before the project begins will let you know what to expect from the inspections. His knowledge of zoning bylaws makes him a valuable first resource when you're looking for land.

Most building inspectors are former builders, though that isn't a strict requirement of the job. Their mandate is to see that the building code is obeyed. Their power lies in the threat of rescinding the builder's license, though they actually do so rarely. If a builder chooses to fight the inspector or ignore the code, the inspector is much more apt to negotiate a solution than to pull a license. His is a getting-along job, and a quarrelsome inspector doesn't last long. In the real world, the building codes demand such minimum standards of construction that your job will barely be affected at all.

Your carpentry crew will probably be siding the house while the trades ready the interior. They may have time to build the garage, the porch, or the deck. They might even leave for another project if work dries up on yours. You might find days when nothing is done. Your best bet in that case is to phone your builder every day or two. He knows he has to get back on your job, but reminding him won't hurt. Don't make threats, but squeaky wheels *do* get greased. You'll preserve your good relationship if your attitude encourages him to return.

11

Siding

SIDING IS THE main skin of your house and has many jobs. It must look good on its own and fit with the architecture of the house. It should protect everything inside it from extremes of weather. It should hold the finish you choose to apply to it. It should endure with aplomb in spite of environmental conditions and some neglect and abuse by humans.

The siding is a major visual component of the house, and the type will be mandated by the architect. If you're designing your own house, you probably squint your eyes and imagine it finished; the siding is part of your image, and thus is self-selecting. You may have some choice of materials within the design, though. You'll be limited when your house fits into a definable architectural period or style, and bewildered when you go for "contemporary." Contemporary seems to mean anything goes.

Colonial- or traditional-style houses usually use horizontal-lap bevel siding, called clapboards from the sound two make when hit together. While early colonists used wide (eight to ten inches), rough boards lapped one over the other, as soon as they could they were moving up to the world of finer, narrower pieces spaced closer together. It's also likely that the settlers used rough vertical-board siding; more on that later.

Clapboards have one big advantage over some other sidings: they are inherently waterproof. As long as clapboards aren't broken, water has to run uphill to leak inside. Even lowly composition-board siding, lapped like clapboards, guides water from one board to the next, and so on to the ground. With vertical siding, by way of contrast, the joints between the many boards are exposed for their entire length. Plywood siding, too, has its joints exposed, though fewer of them.

Another good thing about clapboards is that they will work fine made from thin pieces of wood. This not only conserves trees but subjects the siding material to less seasonal movement and its resultant damage. A thicker board is more likely to contain grain patterns that fight each other and cause warping and splitting and cupping. A thin clapboard that gets wet outside but stays dry inside is much more stable than a thicker board enduring the same treatment. I have seen many houses over 150 years old that have their original clapboard siding intact. Few other products are this durable.

Clapboards were originally regular boards, rectangular in section, usually put on in short lengths. Modern milling has given us the "½" x 6" VG red cedar" clapboard. A clapboard is about half an inch thick at the bottom, a quarter inch where the next one covers its upper edge. The "VG" in the description stands for vertical grain, which means the grain lines are more or less vertical when the board is laid flat. This practice, cutting a board roughly along a radial line out from the center of the tree, produces a stable piece of lumber. Western red cedar is a uniform wood with very straight, fine graining that is soft, supple, and easy to saw into slender long pieces. It has some natural resistance to decay, but can't be exposed to much weathering without some surface protection such as paint or stain.

Western red cedar is expensive, because of steady demand and a dwindling supply. Builders have tried many other woods as clapboards, and some work well. I have seen basswood, fir, white pine, and spruce used with success. Again, the vertical-grain boards are the most stable. The best results come when the siding is painted with real old-fashioned paint to keep the weather away. I would stay away from a brittle wood like spruce, and select only straight and knot-free boards.

It's clear when you examine a very old building that the weak points of the clapboards are where they butt — where the pieces meet end to end and where they meet trim such as corner boards and windows. I aim to buy and install clapboards for the fewest end-to-end butts, since the joints at the trim are a given. I try for very tight joints, to keep water out as long as possible. I urge owners to paint their siding rather than stain it because a good paint film protects the siding from the weather. I understand when they specify stain, an easier finish to renew, and try to make *them* understand they must restain every three or four years. Paint on clapboards will last longer if the boards are back-primed before they're nailed on; see chapter 16.

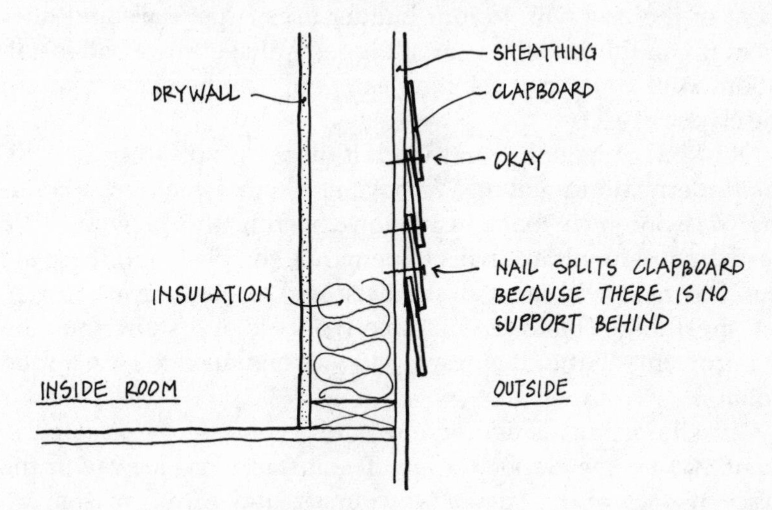

CLAPBOARD NAILING

Nailing is a problem with clapboards. Clapboard manufacturers recommend nailing into the studs, high enough on the clapboard so the nail misses the top of the piece below. The idea is that each board, free to move around a little independent of its neighbors, will have its best chance for a long life. Trouble is, maybe a third of the time you do that, even carefully, you split the clapboard you're fastening, since it's unsupported from behind. The split makes the clapboard's useful life extremely short: only as long as it takes you to rip it off the wall in disgust.

Apple Corps takes a lesson from older houses. Rather than nail each piece with a long nail every two feet into a stud, we use fivepenny hot-dipped galvanized box nails, about eight inches apart, peppering the siding. This approximates pre-stick-framing nailing, with the modern advantage of the galvanizing. Old-timers had to set each nail and put putty over the hole before they painted to prevent the nails from rusting. We don't nail above the lap line, but about half inch from the bottom edge. The myriad tiny nails allow the slight movement of the boards while holding them securely to the building.

Another good system is stainless steel nails carefully driven into the studs. I still like the looks and the slightly flexible attachment of the tiny nail. If your builder uses larger nails and lines them up on the studs, you aren't necessarily getting a bad installation. Still, you might ask him, early on, how he plans to fasten the clapboards.

Our kind of nailing won't work if there isn't plywood sheathing underneath to nail to. We have used our system on a house that has one-inch foam insulation covering the plywood. We used eightpenny Maze nails to penetrate the plywood; these are slim box nails with a small flat head and a rough shank to grip the sheathing. This house is three years old and shows no nailing problems, but that is scant evidence on which to base a solid opinion.

Once the nailing is figured out, care becomes the crucial ingredient of a lasting clapboard job. The installer has leeway in the exact overlap of the successive courses, and within reason can tailor the coursing to produce a neat and tight job. We always shoot for lining up the bottom edge of a clapboard with the bottom and top of the window trim, the top of the door trim, and perhaps the bottom or top of a prominent detail such as a cornice return. This assumes that the finished edges of these features are at equal heights around the building, ensured by accurate rough openings.

The installer must trim off any part of a clapboard that is split or damaged. He must be sure the sheathing is smooth, with no chunky splinters hanging out to strain the thin clapboard. He can cut the pieces with a fine hand saw or an electric miter box or a chop saw. In any case, his goal is a smooth cut and a tight

fit. Rough cuts or loose fits leave a starting point for water
ing its way in.

For years Apple Corps cut all its clapboards with hand
up on the staging, finish-trimming them with block planes as necessary. Chop saws changed that, and we now keep a cutter at ground level, adding an extra person to the installation crew but saving time just the same. And the quite smooth and square machined cuts are more apt to fit each other. I miss the change-up rhythm of sawing and nailing, which provided a better workout, but we only return to it on small jobs.

Less well known in most places than clapboards are shingles, sawed thin from white or red cedar. Shingles take lots of labor to install because they are small and it takes thousands to cover the walls. They last many decades with little or no finish on them, though, and can be cheaper in the long run than clapboards with their traditional repainting. Shingles are great looking on certain oceanside cottages and mansions in the Northeast, where simply nothing else will do.

Shingles can be woven together at a house's corners or butted to a corner board, depending on the design of the house. In either case, we generally fasten a strip of fifteen-pound felt (tar paper) around the corner under the trim or siding, to shed any water that makes its way in. It's all right to install wood shingles over one-half-inch plywood, though five-eighths gives stiffer resistance to the many nails used.

Brick, stone, and stucco sidings are rare in my part of the frost belt, and I've seldom had occasion to use them. The main things to keep in mind with these sidings are moisture control and separating the support for the siding from the structure of the rest of the house. In New England, most masonry sidings are protected by wide overhangs from the thorough wettings that would make them liable to frost damage.

Installation probably causes most of the trouble for T 1-11, that siding-sheathing I can tolerate only on commercial buildings. On multistory walls metal flashing must be continuous between lifts (vertical layers). Anywhere a window, door, mechanical device, or architectural feature penetrates the siding it must be flashed to divert water from the opening. This means having the doors and windows on the site to mark out the exact placement of the flashings. Some kinds of windows won't work unless they're installed *before* the T 1-11, and thus the roofing, a clearly precarious practice. Trim must be flashed so water doesn't sit back there and rot it. Any time you cut into siding, you'd better be sure water can't work its way in. With T 1-11, there's no other layer of anything to keep water out of the structure.

Vertical boards, which T 1-11 tries to imitate, are popular in some sections of the country, particularly on contemporary-style houses. The boards may be pine, cedar, redwood, or some other, usually softwood species. They may be matched (tongue-and-grooved), shiplapped (notched to overlap, one over the next), or battened (separate wood strips nailed over or behind the joints between the boards). They are used sometimes for economy, mostly for style.

Cedar and redwood are stable in the weather but soft. White or ponderosa pine is a little harder but less stable. Yellow pine is tough but quite unstable, sometimes literally twisting itself right off the building. The most common boards in the Northeast are cedar, cheaper than redwood and not quite as endangered a species. We can buy lengths up to twenty feet, speeding installation. Cedar and redwood cut and nail well, hold a finish very well, and look nice natural.

Cedar boards are used for interior paneling, too, and usually come milled for it with matched edges. The tongue-and-groove joint is fine indoors. Outside, it can shrink only about a quarter of an inch before the weather gets in. Because much lumber these days gets shipped before it's fully dry, shrinking is common in the installed piece. I've seen cedar siding boards that have shrunk so much in four years I could get my finger between two adjacent boards.

The shiplapped piece fares a little better because the lap is generally one-half inch wide. The nailing must be very secure, since once the lap opens up, there's nothing between the weather and the sheathing. Cedar is sometimes manufactured with a wide lap known as channel rustic. The wide groove between boards breaks up what's thought to be a boring surface of plain shiplapped boards. This wide lap is more fragile than the full-thickness board, though, and must be well dried and very carefully installed.

Batten siding is probably the crudest variant of the vertical-board genre. As in its origins on cheap outbuildings, boards are nailed up next to one another, then the gaps between boards are covered with a narrow strip of wood, the batten. A trickier installation is reverse board and batten, where the battens are nailed on first and the boards cover them. Either way, you get a thick siding that reacts violently to changes in the weather, moving and cracking and constantly flexing its muscles. What's important on the chicken coop is keeping brother fox out and mother hen in, and who cares if the siding leaks a little? What's important on your house is keeping the weather out of the structure (not to mention the bedroom), to protect your considerable investment in the building.

With any vertical-board siding you must follow several rules. The chief one, which is often ignored, is to install blocking horizontally between the wall studs, every two or three feet up the walls. Clapboards, nailed mainly to the sheathing, get away without blocking because they're thin and don't twist much. Without horizontal blocking it's unusual when vertical boards are nailed into something solid. It's often possible to pull boards thusly fastened off the house with your hands! Putting in all those nailers takes time and costs money, cutting into the financial advantage vertical boards are thought to provide. If your plans show vertical siding, make sure they specify regular horizontal blocking.

If the vertical boards don't reach from the bottom to the top on any wall section, a common occurrence on gable ends, you'll have joints on the wall. Joints between board ends must be mitered to stay dry, a tricky procedure sometimes. Variations in milling or moisture content sometimes produce boards of inconsistent width, so the carpenter has to pick through the pile to find similar pieces. Joints must be nailed into solid blocking, yet shouldn't be lined up across the building, next to each other. Because of all these difficulties, I just don't like vertical-board siding on a house.

There is another kind of reverse board and batten that makes perfect traditional barn siding. This technique employs a full skin of thin (one-half or five-eighths inch) boards with an overlapping layer of one-inch boards on top. The thin boards act as sheathing, making flashing much easier. Since thick boards take the brunt of the weather, this siding installation can last a century or more. And since the barn is open and unfinished inside, the occasional leak matters little. Fairly cheap lumber may be used as the finish layer as long as knots are solid.

When some people say "siding," they mean vinyl or aluminum. Rightly, these are sheet materials, though they mimic clapboards. The installer can't vary the vertical layout of these products to match the house's trim. Aluminum siding is declining in popularity in favor of vinyl, which seems to look good longer. I guess a well-installed vinyl-siding job on a new house will last a few decades if local airborne pollution is kind to it, but who knows? The product hasn't been around in its current formulation long enough to tell.

Clapboard siding is costly to install and maintain. Apple Corps has always shied away from taking on only upscale projects; we like diversity in our customers and in our work. Vinyl is reasonable to install and theoretically requires no maintenance but an occasional washing. But in fifteen years of building we've never installed vinyl or aluminum siding and we hope never to.

Remember my premise that it's our societal duty to use renewable resources where possible, build for the long term, and plan our land use to best benefit the most people. Each house built these days, on its poured concrete foundation, is a permanent part of the landscape, tough to forget about and harder to be rid of if it becomes an eyesore. As an owner, you are responsible for this permanent change. Maybe you don't care if your house lasts a century and are unwilling to sponsor the satisfaction of unborn generations, but I hope you are. I think it a small tax to pay for the privilege of home ownership.

This is a rather lofty argument for wood siding. I know I like clapboards because they resonate with the history of New England. I like the way they smell and the light whippy thinness of them. The trees they're cut from won't survive endless harvesting, though they do grow back, unlike vinyl trees. Installed, clapboards can deteriorate if not tended to occasionally. And they cost a lot. I want you to pay that cost willingly. It helps me to think globally when I buy such things for my house, and I hope it helps you.

12

More Subs

WHILE THE CARPENTERS are nailing on the siding, enter some subs you'll probably never see again, the insulation installers and drywallers. First, however, your builder will duck back into the house for a final check to be sure all the framing is in place. He'll repair anything a plumber cut or an electrician drilled with too much zeal. A thorough cleanup of the whole interior makes the wretched job of the next guys slightly more bearable.

It pays to think of fire protection at this point. Most building codes try to legislate against the easy flow of fire from one level of a building to the next above, and rightly so. Chases must be open at top and bottom to allow ducts, pipes, and wires to be installed, and should then be closed off at top and bottom according to code. I try to imagine the path of a fire starting in the basement or first-floor rooms and trying to get to the roof, and do my best to block those paths with solid wood or masonry. Fire coming up a wall cavity can easily get over the top of a suspended ceiling, for instance, but not if fireblocking (two-by-fours nailed flat between studs) is installed just below the ceiling line. Lots of people die in house fires when a little more time to get out would save them. An hour spent walking around the framed building thinking about flame paths before the subs come in is cheap insurance.

Insulation

The purpose of any insulation is trapping air. It isn't glass fibers that keep your fuel bills down, but the tiny spaces between them. An empty cavity in a building, like an uninsulated void between studs, allows air to circulate. Air moves up the warmer side and down the cooler side and quickly transfers your precious BTU's outdoors. The empty area is an easy target for wind that may force its way in, too. Ideal insulating material would take up little room itself but create many tiny voids filled with dead air.

Most house insulation is fiberglass, fine strands of glass woven into a fabric. In floors and walls it's installed in batts, pieces of common thickness and width that fit between joists and studs. In attic floors it can be chopped up and blown in, filling and covering the framing to form a warm cap on the house. Fiberglass doesn't burn or settle, is relatively cheap, and seems to be inert once installed.

Trapping all the air — between studs, for instance — means filling the whole cavity with insulation. Fiberglass batts are flexible, and work satisfactorily when compressed some, so they can fill odd places well. They don't do that automatically, though, especially around electrical boxes and wires. The open spaces at boxes and wires cause the chill breeze you can feel at a house's outlets in winter. This predicament can be eased in a couple of ways.

One is notching the bottom of all outside-wall studs for the wiring. We make the notch with two saw cuts, leaving a small triangular cutout in the center of the bottom of the stud. The electrician needn't drill the studs for wiring, but can run it along the bottom plate of the wall, then up to each box, so the wiring doesn't push the insulation around much. This wall or its normal cousin can be further helped by just being careful with the insulating. Taking time to fit the fiberglass around the boxes costs you money now but keeps you warmer later. This care is rare to

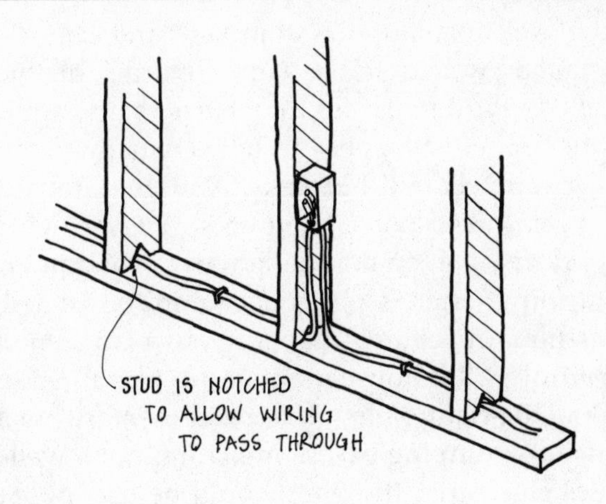

find in installers who are pieceworkers for insulation companies. Ask your builder to hire installers by the hour if he can, or to do it himself.

Other wall insulations are usually blown in after the sheetrock is up. Cellulose is ground-up paper, chemically treated to resist burning. It settles a bit in walls, which eliminates it right there from new houses that can be insulated in other ways. Injected foams shrink a percent or two. Two percent of a fifteen-inch stud cavity is more than a quarter of an inch, space enough for a big breeze. Foams emit a lethal gas when exposed to fire. New products enter the market all the time. I think something that's as hard to replace as insulation should be specified conservatively, so I use fiberglass.

In attics, though, fiberglass is a problem. The fiber size of some chopped products fits in the range of those, like asbestos, associated with respiratory cancers. Since attic insulation is often exposed somewhat, the loose fibers are a health hazard. Cellulose is safer, and a good insulator. I wonder about the long-term effectiveness of the fire-retardant treatment. The settling is no big deal in an attic where you can get plenty of depth. Spraying the installation lightly with water seals the surface to keep dust down. Get your R-44 in the attic from cellulose. And visit your house just after the insulating is finished — it'll never be this eerily quiet inside again.

When the wall insulation is complete and before the sheet-rock is installed, we stretch a vapor retarder on the framing. (They're called retarders rather than barriers now, in recognition of their imperfect effectiveness.) This is usually clear polyethylene, and I specify six-mil thickness. A more durable alternative is Tu-Tuf, a reinforced poly. The film should be tight and cover everything, running in one piece right over windows and electrical boxes for now. There's much debate in the field about using the retarder film on ceilings. Use it if you plan some active mechanical control of the house's humidity, like a heat exchanger. If not, leaving the ceiling film off is a rudimentary but sometimes effective way of dumping excess moisture into a well-ventilated attic. Studies show that the film should be used, but insulation installers and sheetrockers don't like it.

Drywall

When Apple Corps first started in business, we hung and taped our own sheetrock. We got good at it but were slower than the pros, so we began to see the virtues of hiring it done. The first time we tried that was an eye opener.

We were building a handsome architected house out of our home area. We chose a sheetrock contractor from the yellow pages and hired him after a phone call, since he was the only one available. The fellow came to the house to measure for his material order, and was genial and reassuring. He said the "board hangers" would arrive first, and he would be back to tape when they were finished. He said they might look strange but did a fine job.

Late in the morning on their arrival date an old van wheezed up the driveway. What clambered out, leather creaking and chains clanking, was the pride of Satan's Saints, a notorious local motorcycle club. There were four, I think, though I vividly remember only Jules and Frank. Jules was four by four and an Elvis fan, passionate unto violence in his adulation. It was clear we could be maimed for maligning the King. Frank was the foreman, tall and rangy. He bragged for our benefit of bets he had won for accurately throwing his sharpened sheetrock hatchet into certain barroom walls. We were weak.

Their greasy denim vests and DA hairdos notwithstanding, these guys showed up every day and did creditable work hanging the board. It was a tough job, too, with high cathedral ceilings and tricky scaffolding. The boss came only once; he evidently trusted them. His taping job worked out fine and the subcontract was a success. We learned there not to judge the finished product by the practitioners, though we were anxious mothers for a few days.

We've never seen that crew again, but even our current, much more civilized sheetrockers advise you not to "feed the animals." What they really mean is they want to be left alone on the job. They know their pace and style, and the work they do, hold them well apart from most customers. Watching them work is startling. They manhandle heavy, awkward, and fragile sheets into place at speed, and every scratch must be filled in. They endure the body-killing pace that piecework demands, and wash it down with loud acid rock. Small wonder they seek peace in the difference between themselves and mortals.

Sheetrockers come in two flavors, hangers and tapers, under the single subcontract. Hangers install the board, four feet by twelve by one hundred pounds, screwing (never, never nailing) it to the framing with electric guns. When they finish, the mudders or tapers apply paper tape and compound to the joints, fill the screw heads, and make the surfaces ready for the painter. Each crew depends on the other's handiwork, and often blames the other for flaws in the finished product. Hangers do grunt work and tapers need a light touch. Either job is dirty and tough, and you never see an old sheetrocker.

A solid, lasting job is unlikely unless both are done carefully. I think the hanging is more important, because taping problems can be fixed much cheaper. Our favorite crew shoots for fitting the sheets to within an eighth of an inch. Any mistakes leaving gaps of a half inch or more they fill with a tough filler called Durabond. The trade retort to a mention of too-wide gaps between boards is, "That's why they make tape two inches wide." If the board doesn't fit tight, however, the job will be harder to mud and problems will show up quicker.

Good taping is quite artful, as any onlooker can see. The best guys use little mud (joint compound) and leave little to sand smooth at the end. A good taper can make a poor hanging job look good — for a while. At the end of his work, a good sheetrock contractor will ask you if you have any problems with what he's produced. He probably won't mind repairing defects in his work then or later. After a heating season or two is when most difficulties show up. Most delayed sheetrock problems come from the framing, not the board.

Lumber shrinks and twists as it dries and settles under a building's loads. The sheetrock is stable but not strong, so it reveals the gyrations of the framing behind it. Once the building has settled, the sheetrock may need to be rescrewed in a few places, and some joints repaired. This is normal for the product, though minimized by a thoughtful installer. Ask your builder or sheetrock man to see who will be responsible for the touch-up. Otherwise, my reluctant advice is to stay out of the sheetrocker's way.

Another surface product making gains in housing is veneer plaster, or blueboard. This stuff is much like sheetrock, but the joints are taped differently, and the whole surface of the sheet is skim-coated. It makes a better finished surface, but one that takes more skill to produce. It used to cost more than sheetrock, though now the drywall manufacturers are cutting quality and raising prices, and blueboard looks better and better.

The final step in the sheetrock installation is sanding the mud, a task normally reserved for the crew's rookie. The proper approach to sanding is going around the entire surface with a strong light held right against it. Shadows highlight an irregular surface. Careful sanding cuts down the mounds and smooths the trowel marks. A good sanding is crucial on walls and ceilings slated for surface-mounted light fixtures. The low-angle light they cast emphasizes peaks and valleys.

The only way to sand sheetrock is to get your face right up to it and rub. The hapless novice gets covered inside and out with fine dust. So does your house, of course. This is the dirtiest your house should ever get, and the certain nadir of any remodeling job. The sheetrock sub and the general usually fight to get the other to clean up the scraps and dust. Whoever loses should clean the house with mops and vacuums, and the rookie with beer. The next phase of the job, the finish work, is set to begin, but first let's check back with the carpenters working outside.

13

Porches, Decks, and Garages

A PORCH OR DECK gives you another room outdoors, a level place for your bridge table or sleeping bag. Being out on the deck is different from being out on the lawn, though. The edges of this platform, like the walls of a room, contain you and your activities within familiar and manageable bounds. A Victorian screened porch and a ground-level California deck share this quality of taming the elements to your use.

If you plan a regular entrance to your house across the porch or deck, make sure it doesn't cut into the deck space you want. You'll need a clear swath four feet wide for a principal entrance. A screened porch plan means you'll have to open only one door at a time when entering, a convenience when your arms are full. A door with a screen or storm door attached is a tricky obstacle with no free hands.

Make sure you plan enough room on your deck or porch for the way you will actually use it. If you'll have a table out there, a ten-foot width is minimum, twelve is better. Don't have this deck open into the living room if you envision the Tonka operators crowd meeting just outside. If you think of partying and barbecues on the porch, leave plenty of room around each activity area. Remember, these outdoor square feet are six or eight times cheaper than your average interior ones, and will get lots of use. A porch is good for many things besides a rocker.

Once you plan for people actually using your new porch or deck, see if you can fit its style into that of your house. Here decks and porches part company, for both don't often suit the same style of house. The details of the design will either integrate one or the other into the house's lines or obtrusively call attention to itself. Bold new themes are something to explore in your remote garden shed, but fit the porch to the house.

Location in the façade is the most important criterion. Make sure a major valley in your roof doesn't discharge right overhead. The low-pitched roof of a traditional porch will easily build up with snow, so don't burden it with the refuse of higher roofs. A porch needn't sit against the house symmetrically, and indeed gets more air if it wraps around a corner. One's eye sees the porch as an adjunct to the main building, and it helps if the visual weight of the porch is balanced by some mass of the house proper.

Sun, breezes, views, bugs, privacy, access, space, and style all will matter to you in the finished product. As long as you build carefully, this will be a lasting structure. Who knows? You may even close it in someday for a family room, as many have done to theirs. So start from the bottom and make it sturdy, then use it as you like.

Recall the foundation section of chapter 5. Any structure attached permanently to your house should be supported as sturdily as the house itself. The points of attachment will suffer if settling or frost heaving moves the porch but not the house. If your house gets a poured concrete foundation, so should your porch, stoop, or stairs. The two most common repair jobs on porches and decks are repairing rot and curing underbuilt foundations.

Most porch and deck foundations are concrete poured in round columns such as Sonotubes. These work fine, provided they are poured on top of a footing with a piece of reinforcing rod sticking out. Large, flat rocks make decent footings for light decks in well-drained soils, but lack the re-bar pin and its solid connection. Follow the guidelines in chapter 5 for frost depth, tamping, and backfilling.

If the deck or porch framing is to be almost directly on the ground, I pour the tubes right up to the bottom of the framing. Otherwise, I often pour them to finish grade, set an anchor bolt in the wet concrete, and go up from there with wood posts. If a customer or architect insists, I will even keep the tubes slightly below grade so only the post shows. It's no huge job to replace the post section after a few decades when the rest of the trim, lattices, and skirts are rotted, too. It's always tempting to eliminate the hassles of pouring concrete by burying the treated posts down to the footings, since the posts are supposed to last thirty years or more. It's often very difficult to replace them if they do cause trouble, though, and prudence rules in favor of concrete.

Any outdoor stairs deserve a solid footing below frost depth. A short and simple two- or three-step stairway can be built on treated posts buried in the ground because it won't cost that much to replace when the posts finally give out. Wide, tall, or ornate stairs, however, should get concrete to grade level and pressure-treated framing above grade.

I use only pressure-treated lumber for all outdoor framing. Decks, though they let water pass right through, trap enough to rot the joints in eight to ten years. I avoid a deck frame design that calls for joists or girders to be doubled up for strength. The joint between pieces traps moisture and guarantees rot. Any place I double the framing I use spacers to separate the pieces an inch or so to allow air between them. I figure you can replace the deck surface every decade if you wish, but the framing ought to stay intact far longer. This technique should be noted on the plans.

Outdoor floor frames can get dangerously weak before anyone notices, although sometimes the railings and decking boards rot at the same time, alerting you to trouble. In my part of the country an untreated deck frame, no matter whether it's stained or slopped with preservatives, rarely lasts beyond ten years.

Commercial joist hangers have no place under an outdoor deck. Most are galvanized very lightly, and I've seen hangers rusted through in twenty years. I'd rather see joists resting on a ledger, spiked to the girders or to the house through spacers. If joist hangers were to improve by more galvanizing, I'd recommend them.

Recent studies have questioned the safety of pressure-treated lumber as a deck surface because of the toxicity of the treatment chemicals. The evidence is inconclusive, but many inquiries about environmental toxins turn up problems worse than the fears of the early researchers. (Pressure-treated decking is also hard to paint; more about this in chapter 16.) Your architect should discuss the subject with you and let you choose. Substitutes for pressure-treated decking are redwood and cedar, with redwood being more expensive and longer lasting. Most other commonly available woods just rot too fast to be worth putting down, at least in a humid climate.

I like fir flooring on a porch with a roof that can be expected to keep most of the water away. This traditional narrow-strip tongue-and-groove floor has changed little for decades, except the lumber has gotten somewhat thinner. Fir is tough and wears well and is fairly easy to paint. There are other options. Cypress is tough but splintery. Locust is very tough and hard and rot resistant, but hard to find and prone to leaping out at you as you walk by on a hot day. Fir rots eventually, but I haven't found a better all-around product.

Porch floors must be framed with a slight pitch away from the house, usually one inch in ten or twelve feet. This lets water run off, albeit slowly, thus preserving the flooring and the house. A steeper pitch would drain better but be uncomfortable to walk up. The floorboards should lie along the slope, ends toward the house. If the porch goes around a corner of the house, the flooring should make the turn and end up sloping away again on the adjacent wall. The floorboards at the turn can make a pretty pattern on this part of the floor.

Traditional porch roofs are framed just like hipped house roofs, but usually flatter. A hip roof gives the advantage to the design of a level sight line all the way around the porch, emphasizing the horizontal aspects of what might be a boxy shape. Historically, porches were added to houses, and a nice-looking porch can have a roof line much different from that of the house. Southern houses were more likely to have a porch built right into the design, often under the shelter of the main house roof. I know more about New England porches, which are enjoying a second heyday as our colonials, Capes, and Victorians reseed themselves across the country.

Your porch should get roughly the same cornice details as the house, but reworked for the flatter pitch. An eight- to sixteen-inch overhang is traditional. The wider dimension protects the deck, flooring, and railing better, at the expense of some trimness of appearance. I always make sure to ventilate a porch roof just as I would a house's, to save the roofing. Porch columns require ventilating too, if they're hollow, and this is not optional if you want the paint to stay on them.

On solid posts I sometimes drill a quarter-inch hole into the house side of the post near the bottom, angled downward. A little wood preservative or a linseed oil and turpentine mixture squirted into the hole from time to time will protect this vital area from rot. An existing porch can benefit from this treatment, too.

Old-timers used tin sheets with the edges soldered and folded over, making a seamless porch roof. These can last over a hundred years if ventilated underneath and painted before any rust shows. Tin roofs are rare now because they cost so much to install. If you're rebuilding an old house and have a sound tin roof, by all means save it.

The inexpensive modern version is double-coverage roll roofing. This is heavyweight asphalt shingle material that comes on a roll. "Double coverage" means each sheet is covered halfway by the next above, and covers halfway the next below, too, so every spot on the roof has a double thickness of roofing. The trouble with this material is the large pieces, which move around in the sun and cold. The stuff gets brittle as it ages, and the moving causes cracks to develop. Eight or ten years is the most to expect from roll roofing.

Rib-style metal roofing may be used on a low-pitched porch roof but will leak a little over the years as the material slowly works the nail holes larger. Terne metal with its folded seams and hidden nails works fine, and is the modern equivalent of soldered tin. Wood, tile, and slate all suffer the same shortcomings on shallow pitches — the water must be running to run off.

I usually recommend using standard asphalt roofing shingles, whatever is on the main house roof. So that these won't leak, I install a continuous waterproof membrane over the whole roof deck first (see chapter 8). When the time comes, in twenty years or so, for a new layer of shingles, the membrane will seal its nails, too. After that it's time for a new one, but that's at least forty years from now.

Once the roof is on, it's time to install a railing, stairs, trim around the deck, lattice below, and maybe a few fancy brackets or moldings. You can specify pieces from your local lumber yard or millwork shop, and several mail-order suppliers of reproduction millwork sell the stuff. The main thing to remember with any of these items is that they'll get wet, dry off, and get wet again many times over their life. They must be designed to shed water and be kept caulked and painted or they just won't outlast the mortgage. A fancy porch deserves the best paint job on the house.

In my area, any deck over thirty inches from the ground must have a railing, and the railing must not allow a six-inch sphere to pass through at any point. This rule saves babies and is a good standard no matter where you live or what your current family makeup.

This code also limits the designer. If the deck is simple, the rails should be, too. Built-in seats, planters, and gas grills are fine if you have actually used a similar setup and know that's what you'd like. To me the most usable deck plan is one that sacrifices fancy details to gain more size. Would you rather perch upright on a wooden bench or stretch out in a folding chair?

Any way you design it, build it safe, and the higher off the ground the more careful you must be. Tall deck posts must be stout and cross-braced so the whole thing doesn't rock and roll when it's loaded with dancing partygoers. Just overbuild the heck out of it, as my partner Ned would say, and everything else will take care of itself. Stairs, posts, railings, flooring — all must be super sturdy.

Your design may include masonry stairs, of brick or flagstone, or concrete. These should have a foundation all their own, poured at the same time as the house foundation. These stairs are massively heavy, and if they are set on the backfill of the house foundation, they are bound to settle away from the door. Concrete and masonry steps have rotted many a sill, too, and here's why.

A standard riser, the vertical distance from one step to another, is around seven inches. Normally, the first-floor deck of a house is a foot or so above the foundation. To stick with the seven-inch riser, the builder must place the top step at least four inches up on the floor framing of the house. Concrete and masonry stairs stay soaked with water for long periods, hold that moisture against the house, and rot it.

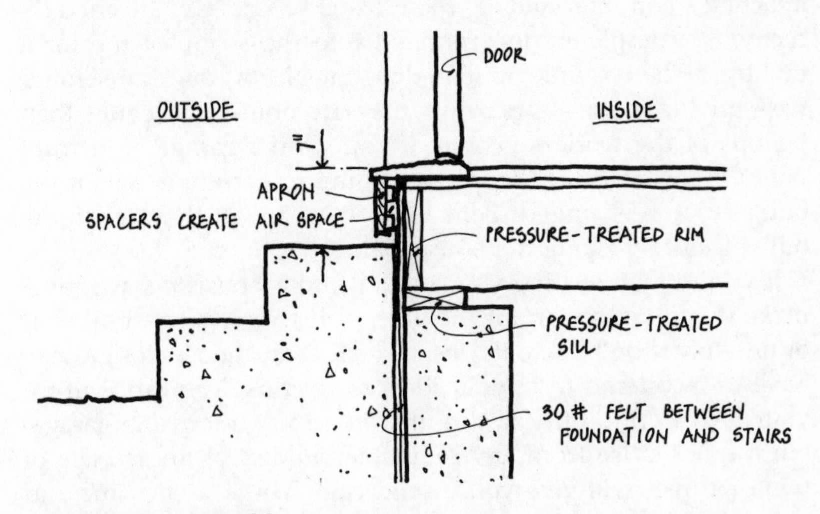

MASONRY STAIRS

The solution to this problem should show in your specs and on a section drawing through the stairs. Specify a pressure-treated rim joist in the area of the stairs. Show thirty-pound felt between the concrete or masonry and the plywood. The apron board under the door's threshold should be spaced away from the wall a quarter inch or so for air circulation. This apron should come down almost to the top step, though not touch it, to keep debris from between house and stairs. Finally, slope the steps slightly away from the house. None of these precautions will last if the stairs are on a weak foundation and move around.

An inexpensive alternative to these woes is an open wooden stair, set on a flat rock or two over some good coarse backfill. This stair should be attached to the house somehow so it can rise and fall a little without harming the house or the attachment itself. I've used barn door hinges, a steel rod pivot, and simple steel angle irons with loose lag screws, all with lasting results. None of these is ideal, but does save the digging and pouring and expense of fancier stairs.

The garage

Your house plans may call for a garage, too. It may be detached, attached to an intermediate room like a breezeway, attached directly, or completely incorporated into the shape of the main building. There is little point in debating North Americans' fondness for their cars — we know they are housed far better than billions of the world's people. If you want a garage, you want one. Often the garage is the first thing to get the ax when the budget battle is waged. Sometimes it's a tough thing to throw out without wrecking the house plan.

If you're just wishing you could afford a house *and* a garage, make it simple to wait for the garage. Plan a separate building, or one that won't make its absence felt. Your house design may have to be altered from your ideal to get this. You can wait for your bids to find out for sure that you can't afford the garage, but a quick price from any reputable builder, or an average of two or three, will give you a solid clue. You'll waste time and money designing a house with a garage and then chopping off the garage.

It's often a mistake to pour the garage's foundation now and plan to build it "in five years or so, when we have some money." The trouble is, that foundation will sit there for more years than you first imagine. The anchor bolts must be sawed off (or left out) for safety. And who knows what you'll want in five or ten years? Maybe the garage should go somewhere else entirely? You might even move, and the foundation will not help the house sell. And it will be in the way constantly from now until whenever.

Builders have for years put garages in the basement of houses. It makes sense. The foundation is already there, the concrete floor is there, and having the car inside is very handy in lousy weather. Adding a basement garage makes the whole package more valuable without costing much. For a separate two-car garage you'll pay a builder ten thousand dollars or more, while one in the basement can be had for two.

I think it's a great idea if you can work out the design. I wouldn't raise a house six extra feet out of the ground just to put the garage under it. I wouldn't slope a long driveway down to the cellar level from an otherwise level lot, practically guaranteeing flooding problems. If your lot slopes enough to make driving into the basement practical, great. Arrange the grading for an outside area that slopes down from the garage-floor level so you don't get ice ponds.

The car mustn't put your house in danger. Most building codes call for fire protection between garage and living quarters. This usually means taped and finished fire-code sheetrock on walls and ceiling and a one-hour fire-rated door into the house. Be sure the finished surfaces present no channel, however small, for flames or fumes to get into the rooms above or into the structure of the house. Outlets and switches, garage door brackets, and door trim must fit tight. Fiberglass insulation in all the cavities around the garage makes sense thermally and acoustically and also can help contain a fire for a short time. Electric smoke detectors connected to the house wiring are a must.

More likely you'll design an attached garage at grade level. You can choose one that is incorporated under the main house roof or one with its own roof line, to make it appear distinct. Fire safety rules are the same as for basement garages, but make sure you're thinking of all the possible flame paths, including through the roof structure. Local codes often cover this subject, but think about it anyway.

A standard-size two-car garage is twenty-four feet square, measured outside the building. Many are built smaller, down to twenty feet square, which is all right for compact cars but hell for my pickup. Actually, the pickup will fit inside, but there's no room left to get in and out and store the trash cans and mowers and sleds and all. Some garages are never used for cars, simply because they didn't include those few extra feet for the junk. One mower or wheelbarrow can keep the car out. Inexorably, the extras take over until redeeming your parking space isn't worth the struggle.

You'll have a choice with doors, too. The standard size is nine by seven feet (width named first, as usual). If your cars are small and your driveway goes straight into the garage with no turns, eight feet is enough width, but only barely. Drivers just learning to guide the ark will appreciate a nine-foot opening, and any turn near the door makes that size mandatory. If you use a roof rack regularly on a full-size car, or drive a truck or van, an eight-foot-high door makes sense also.

I'm a great believer in remote-control garage door openers, which, in an attached garage, let you into the house without stepping outside. Buy a middle-of-the-road model to get the sturdier hardware and a timed light. Have the electrician wire an outlet near each motor unit, and make sure it's all included in the contract.

The garage should be built roughly the same way as the house, with pressure-treated sills at least eight inches off the ground and with solid framing. Siding and trim to match the house is proper, and the doors should make an attempt at similar styling. Any poured concrete apron outside the garage doors should be poured at the same time as the floor and secured to it with rebar. Anything wider than about one foot will likely separate from the slab anyway, so make the apron narrow.

Roof the garage just like the house, and be sure to provide for ventilation. Trussed rafters get you a flat ceiling and no attic storage. With a site-built roof system you can get some storage overhead, but you'll need central posts and a girder under anything heavy, like lumber, books, or furniture. Include at least a couple of windows for light, especially if you don't plan on finishing the inside of the garage. Don't hope the windows will ventilate the roof; provide something like ridge and eave vents that work all the time.

Floor drains are optional, depending on your local climate. Building codes may specify where these drains lead, as they are a possible source of pollution if motor oil or other toxics get into them. Don't tie them into your house's sewage system. Particularly in a basement garage, if your floor drain is piped into your perimeter foundation drain, will the floor flood if the pipe outlet gets plugged? Think about where all the water goes, and where it might end up if everything doesn't flow as it should. It might just be easier to sweep the water out of your garage occasionally.

The garage should be big enough to be useful. It should be well integrated into the house design in both shape and detail. It should be built as carefully as the house, especially if it is attached to it. And, at ten to twenty thousand dollars, it should fit into the budget early on or be left out of the plan.

Part Three

Finish Work

14

Stepping Back Inside

IF I'VE FIGURED my work schedule properly, my crew is ready to move to the finish work inside right about when the sheetrock crew leaves. The house is a dusty mess, so we spend a while cleaning up, sweeping thoroughly and vacuuming everything we can. From here on in we'll dance a constant two-step of work and cleanup, trying to strike a balance between productivity and order. Some carpenters — my partner Rich Gougeon, for example — can produce a marvelous piece of goods amid chaos. Others, like me, spend time cleaning up so they can work without distraction. You should be able to gather from visiting work sites and talking to his previous customers the general conditions your builder will tolerate. If a mess drives *you* to distraction, remember that when choosing a builder. The finished product, though, is what counts.

Floors

Often the first finish work we do is installing the finish wood floors. It's great to cover over that last haze of sheetrock dust and have a uniform surface to work from. Fitting flooring around door jambs is a painstaking job we can avoid by installing the doors after the flooring. Old-timers rarely did this, but sanding a floor was much harder long ago, so floors often went in last so they'd stay clean. In the plaster days, before World War II, floors were about the last thing to go in, even including baseboards. If for some reason we have to wait to put down the wood flooring, we'll slip a scrap of flooring under each door jamb when we hang the door. Then when it's flooring time, we can slip the temporaries out and slide the flooring under the jambs, skipping the tedious fitting.

We put in the wood floors before any other type because it's easier to fit the other materials to the wood than vice versa. Carpet, vinyl, even tile, is more adaptable than wood. We aim to have the finished floor surfaces flush with one another, and vary the underlayment thicknesses to make it so. Small differences in height are guaranteed trippers.

Old-timers laid rosin paper over the subfloor not to prevent squeaks as some surmise but to keep dust, fumes, and drafts from filtering up from the cellar into the rooms above. Paper is unnecessary over sound plywood subfloors. Floors squeak because a floorboard nail is loose and rubbing against the subfloor or the board itself. How can one board lying on another produce a squeak when you walk on them? To prove that nails cause squeaks, pull one out of a noisy floor and you will inevitably find that it's shiny from the friction. Threaded or serrated flooring nails stay put as long as the subfloor is flat and sound. Tapered cut nails, the old-timers' favorite, don't hold floorboards well at all.

If the floors are to be different in adjacent rooms, we make sure the change will occur under the door itself, so each room will look complete when the door is closed. If the ends of wood flooring boards will show in a doorway or archway, we often stop them short and nail a strip of flooring crossways at the ends for a more finished look. We experiment with interesting patterns where the room calls for it, around hearths and other architectural features. We mix short and long strips to avoid bunching all the short ones together at the far edge of the room. And we sight down every piece for straightness, cutting up the barrel staves for ends or kindling.

The commonest wood flooring is strip, about 2¼ inches wide

in random lengths. Strip flooring bears a reasonable price because it can be made from small and irregular pieces of the tree. In New England red oak is very popular, with rock maple and white oak rarer but desirable. Red oak carries the lowest price tag, though there isn't much difference between the three. White oak is hard to find, tougher and browner than red, with a more uniform grain. Maple is very hard, gymnasium floor stuff, and light-toned. I love it.

There are less expensive woods. Pine strip flooring, usually around three inches wide, is of southern yellow pine — stringy, unruly wood given to twisting but quite durable when fastened down. Pine has wild and distinct grain patterns some people like, and it fades to a yellow-orange, luminous in the sun. Yellow pine is available in wider boards also, though in that configuration its uncooperative nature comes on strong. Fir porch flooring can be used indoors, too. It has very even, straight grain lines and a nice reddish cast under varnish. It must be laid carefully, though, since it can be splintery where the grain runs out the side of the board. Probably the best-looking floor I've ever seen was laid of cherry strips, custom made by a flooring mill.

We're careful to start the runs of strip flooring straight, to ensure that they march evenly across the room. We use plenty of nails from the floor nailer, a hammer-driven machine that tightens the strip and nails it all at once. Sometimes Apple Corps hires a specialist to install our floors, but not often, because we think we do a more careful job. Running the nailer is hard on the back, so we old dudes lay out the pieces for the young guys to nail down. It's a kind of violent release, using a nailer, well suited to the frustrations of apprenticeship.

Another type of wood flooring is wide pine boards. This is white pine, silky soft between the knots, docile even at great widths, and soon reminiscent of the worn floors of Colonial days. Seasoned properly, this lumber will lay and stay tight through its life. Its softness makes it somewhat unsuitable for entranceways and staircases, but lovely in a living room. Furniture and accidents will mar it, though to those who like it, these marks add character, not shabbiness. We often screw down the wide boards into the joists below and plug the screw holes with wood for a nice effect.

Hardwoods of various species are available in wide boards, oak chief among them. Oak is available in prefinished "planks" with contrasting plugs emphasizing the fasteners. Oak is immensely strong and tough, and the prefinishing helps the boards stay flat on the floor by keeping moisture migration at bay. Really any board can be used for flooring, though the common ones have the best track record for good looks, long wear, and predictable performance. I've even wrestled into place an entranceway floor of black locust, rotproof and tough enough for centuries.

Finishing the wood floors comes at the end of the whole job. This procedure involves cleaning out the rooms, sanding and edging the floor, filling any imperfections, buff-sanding, sometimes staining, and applying several coats of finish. I try to allow a week for the procedure, including a couple of days for the finish to dry hard before it gets traffic. This is the week when the owners are anxious to move in, but sanding transforms the floor so much, I can usually get patience from the most eager customer.

Sanding and finishing are often a subcontracted job, available in my area for around $1.30 per square foot. It may cost Apple Corps twice that to do it ourselves, and we often do it anyway. It is also something lots of do-it-yourselfers try once — the uneven results are evident nationwide. Most people who have refinished their floors are only too happy to pay someone the next time. It's hard on the body and ears and lungs, and harder still with commonly available rental machinery.

Apple Corps uses a huge old American Eight, stable as a truck and almost as heavy when it's time to lug it upstairs. Its weight makes for a smooth trip across the floor, though, and less work for the operator than trying to control a jittery modern version. We follow each successively finer grit sanding with a pass around the perimeter with an edger, the devil's own machine but irreplaceable at its task. When the final edging is done, we clean out the corners with a pull scraper, a regular paint scraper kept scrupulously sharp with a file.

Next comes a once-over with a buffer fitted with a sanding screen. This is a fourteen-inch-diameter slow-turning wheel under a heavy motor, and the sanding is done with a carbide-coated 100-grit sanding screen. This disking serves to merge smoothly the sander and edger digs and to take off the grain feathers one-way sanding might have raised. Going over the scraper marks in the corners with an orbital sander blends them in nicely. We sweep, then vacuum the floor carefully, making sure window sills, electric outlets, and baseboard radiators are free of dust that might fall onto the wet finish. On a showy floor we pick up any remaining dust with a tack cloth. This whole paragraph describes operations that contract floor sanders commonly leave out. I suggest you risk a contractual transgression into technique here, and specify buff-sanding before each coat and a brush finish.

I almost always use polyurethane coatings on floors, for their wearing abilities. The first two coats should be the tougher gloss finish, and the third (or fourth) and final can be satin if you want. After each but the last coat I sand the corners by hand and run the buffer quickly over the floor again to smooth down bubbles and dust mountains with 120-grit screens. I vacuum and tack the floor as before, then brush on each coat evenly. Faster lamb's-wool applicators tend to leave many bubbles, which harden as they break the surface, leaving it rough to the touch. The final coat should be left alone at least forty-eight hours, and the house temperature throughout the process should be around seventy.

I will stain a floor if the customer insists, but I usually can talk him out of it. It's very hard to get stain to be absorbed evenly into flooring, and blotchiness looks lousy. Areas that are tough to sand — underneath overhanging built-ins, for instance — may show rogue sanding marks brilliantly dramatized in stain. Stains also should be chemically compatible with the finish material you select.

I sometimes use quick-drying urethane when time is a big factor. The high percentage of volatile carriers in this product means you're getting less of the solids, the tough body of the finish. In bedrooms, though, where the finish will never wear out anyway, I sometimes substitute the quick-drying stuff to cut down the wait and get two coats on in one day. I wouldn't use it in any moderate- to high-wear area. And I don't use any store-brand coatings. Sanding floors is hard, and I'd hate to risk three days of it to save fifty dollars on urethane. Zip-Guard, Benwood

Finish, and McCloskey's all flow nicely, level and wear well, and have a high percentage of solids.

There are other floor finishes worth mentioning. Linseed oil, cut with turpentine for the first coat or two, makes a dull finish with none of the plastic qualities of poly. Boiled oil is the only one to use, and even then this finish sometimes stays sticky (and stinky) for days or weeks. Tung oil finishes are similar though less sticky, and some say mixing tung oil with polyurethane gives it a nicer tone. Some people are strongly allergic to liquid tung oil. Spar varnishes are great outdoors and good in, though not as tough as poly. In fact, little else is as tough or as simple, and those are the qualities you want in your floors.

In any case, most wood floors are a lifetime proposition for a house. Unless everyday traffic brings in lots of dirt and water, a wood floor will outlast generations of kids, dogs, and friends. Even a softwood floor getting little care will last decades, though it won't look new very long. A paved walk outside and a doormat will help a hardwood floor look great for many years. Consider your wishes for the life of the house when you specify the flooring.

I try to schedule the subs so the painter can get most of his work done before the floors are finished. Then, when everyone's done marking up the walls and baseboards, he can come back for the final coat or touch-up. Carpeting, floor tile, and vinyl flooring go in at the very end of the job, so they're brand new for you. You've already read about wood floors and sanding them. Here are a few thoughts on some other options.

Tile floors can be tough enough for the heavy traffic of a restaurant or airport and will last forever in a house. Rarely do they, though, because new owners often tear out tilework. Kitchens and baths, especially, seem to be changed with each move. I've torn out much more tile than hardwood flooring in remodeling jobs. Luckily, most of it wasn't laid too well.

Look ahead to chapter 15 for a survey of tiles' features and flaws. If you decide on tile, follow some basic rules. Specify extra-rigid framing for the floor under the tile: double joists or regular joists but on twelve-inch centers. Instead of a half-inch subfloor, use the more rigid three-quarter-inch sheet. Any movement at all in the underlayment spells ruin to a tile floor, which is really more compatible with concrete bases than with wood.

Tile over clean, solid, wood underlayments should be laid in thinset mortar mixed with a latex additive in place of water, which makes an incredibly strong and ever so slightly flexible bond. This bit of give is what makes it outlast mastics on wood. There are good mastics — Fuller's Double-Duty, for one — and I use them where the tile won't be wet too often. Thinset is better, though more trouble to use, since it must be mixed in batches and each lasts only an hour or so in the pan.

Mastics got a bad name years ago when they started to replace cement mortar as the setting agent for tile. Mortar is tricky to work with and must be a couple of inches thick to be strong. In a house where that thickness would mean a step in the floor height or a special framing job, mastic seemed like a better deal. Trouble is, much of it dried hard, and tiles popped loose when the underlayment moved slightly. Moisture got through the mastic into the underlayment and swelled it, cracking the grout and letting more moisture in. Thinset holds just fine even under water, and today's glues make underlayments more stable than ever.

Carpeting and vinyl flooring are often supplied by the same subcontractor. Regardless, the style and color and price range are things you're commonly left to decide on when you can see the rooms take shape. Write your contract with an allowance of a fixed dollar amount per unit of material. I suggest starting with twenty-five dollars per yard of vinyl or carpet, and eight dollars per square foot of tile. You can choose the products you want knowing how much you have to spend. If you decide on a more expensive flooring, you will pay the builder the difference between the allowed amount and the actual price. If you choose to economize, the allowance lets you net the difference.

Interior doors

After the wood floors and underlayments are fastened in place comes the installation of the interior doors. Doors are expensive, large, and important to the style of the house. Using the wrong ones can be a curse. There's a huge variety of door types and styles just among the factory-mades, and custom-made doors go from there to the moon. You should understand a few general things about doors so you can decide what to specify.

Door types include swinging, sliding, pocket, bifold, French, and folding. Swinging doors are regular old doors, hung on hinges. A variant is the double-swinging door often used between kitchen and dining room, or as café doors. Sliding doors show up most often on closets. They are inexpensive and will allow you into only half the closet at once. You get more access with bifolds, but at a price. Pocket doors slide into a recess in a partition, helpful when a swinging door would intrude into some vital floor space. The rule with pocket doors is to buy the best hardware you can. French doors are just pairs of swinging doors. As a rule they are mostly glass in a wood frame and are used to separate one formal area from another. The folding door is like a many-layered bifold, and accordions over to the side and out of the way, albeit somewhat awkwardly.

The simplest door style is flush, which means flat and plain. Flush doors are available with hollow and solid cores. Hollow-core doors depend on folded cardboard, inserted between the narrow wood frame rails, to fasten the thin wood skins to each other. There is a block of something solid where the lockset is to go, but that's the only solid part of the thing. A solid-core door is filled with coarse particleboard. It is somewhat stronger, much tougher when kicked, and many times heavier than its hollow cousin. The wood skins come in a variety of species, starting with lauan, a variety of Philippine mahogany, going through birch, the best grade for painting, and up to oak, ash, and a few others at the top end of the line.

Hollow-core doors are truly cheap, and really suitable only for closets or maybe pocket setups. They give no satisfaction in use because they don't act like doors — more like imitations — mainly because they're so light. A solid-core door, though, really can shut things in or out, as a door is supposed to. Solid-core doors are heavier than their expensive solid-wood relatives, and should not be slammed. Flush doors look out of place in a traditional house, but fine in a modern one. They can be had with window or louvers in them for special circumstances, and with fire ratings for furnace rooms and the like.

Panel doors are the alternative. Cheapest is the pressed hardboard unit, which does a pretty fair imitation of a wood door from a little distance. It isn't much more than a hollow-core door with the surface made to look like panels, but is much cheaper than wood. Otherwise you're stuck with a solid-wood panel door. This is rarely made from solid wood, but from smaller, hence cheaper, pieces glued together and skinned over with a fancier wood veneer. They still pass for fancy. Old-time doors were solid wood, and about as apt to warp and stick as their modern counterparts. There are many panel arrangements available, and your architect or designer will have a preference. In general, if a door is classified as being stain-grade, you can assume the surface veneer is carefully chosen and well-joined.

If you cannot find stock doors to your liking, you can have some made. The cost is likely to be at least double, the wait endless, and the quality variable. Very good custom doors are available, and you can get bad ones for the same price. Try to steer away from them if you can. Many custom doors are works of art, and you may be willing to sacrifice a little practicality for that. It's an expensive and sometimes beautiful way to go.

You may have leeway in your plan on the size of the doors. I recommend having no room door narrower than thirty inches — two-six in my feet-and-inches language. A tiny room can get away with smaller, but try to avoid such crowding in your design. A two-eight door feels commodious by comparison, and from this size up you start having to move your body to move the door. Three-oh doors are big, and look best in formal rooms. Eighty inches, or six-eight, the standard door height, is fine for most uses. A six-six unit, the other standard height, might better fit the low ceilings of a Cape's second floor.

A swinging door is described as left- or right-handed, depending on the direction of the swing. The best way to remember which is which is to stand in the doorway, back your backside up against the frame with the hinges on it, and pretend your arm is the door. With butt (yours) to butts (the name for door hinges), a right-hand door swings to your right. Be careful, though. A few manufacturers, Raynor Door for one, uses "right-hinge'' in place of ''left-hand,'' the reverse of my system. That's fine, as long as each person knows what the other is talking about.

Doors are also classified by the type of lockset they will get. A ''passage set'' is a regular room-to-room doorknob; a ''privacy set'' locks from one side; a ''bath set'' often has a different finish, such as chrome, on the locking side. Closets may have dummy knobs that mimic the room doorknobs but have no works in them, and rely on a magnetic catch to hold the door shut. Colonial-reproduction hardware is available with the thumb latch on the side toward you or the side away, depending how you order it. And don't be too thrifty on interior locksets, as the cheapest ones won't last long at all. You needn't opt for the most expensive you find, but stay away from the blister-pack critters in the local discount store.

I always specify three butt hinges for each interior door. Cheaper doors come with two, but that isn't enough. Solid-core doors that get lots of use should have 4-inch hinges instead of the standard 3½. Be sure to order the hinges in the hardware finish (color and surface quality) you'll use in the rest of the house. An exception, a bathroom door that swings into that room, might have chrome hinges if the rest of the room features chrome accessories. The visible part of butt hinges, the knuckles, show in the room the door swings into.

I can order the swinging doors prehung, as units hinged in their frames, or jambs. Prehanging saves us time on the job, and the supplier does it cheaper than we could, and generally as well. If we're using special jamb lumber or fancy or unusual hinges, we'll have to hang our own. We usually order prehung, uncased units. (Casings are the trim that shows in the rooms, sides and top.)

If the casings are already applied to the door unit, the jambs have to be made from two separate pieces so we can install them from each side of the opening. These units are called split jambs. Split-jamb units go in easily but are the devil to make sturdy. Because the casings are in the way, it's tough to shim the jambs into alignment. The manufacturer instructs the installer to use shims but knows it is rarely done. Most times precased doors are nailed to the house only through the casings. With feather-light hollow-core doors this setup might last a few years, but then everything starts to shuck around. Precased doors are out on all but the least expensive jobs.

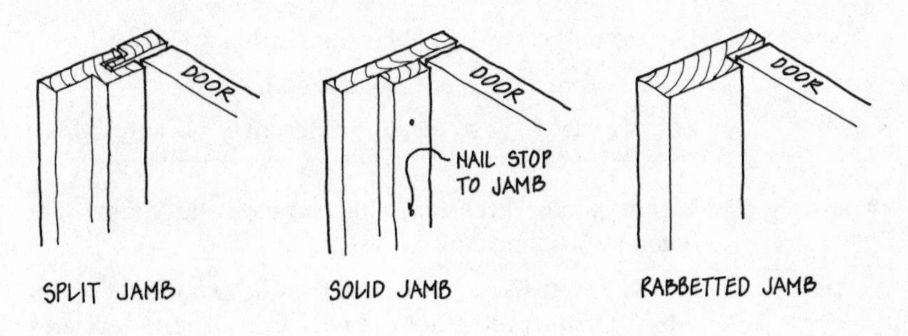

DOOR JAMBS

A variant is the solid jamb, the one I like to specify. This means simply that the door jamb is a solid board with a door stop (the wood strip the door hits when it's closed) attached. On a split-jamb door the stop is part of one of the jambs and the other jamb slips behind it. An even more solid solid jamb is the rabbeted jamb. This is made from a thicker piece of lumber, and the stop is part of the board. Rabbeted jambs are standard for exterior doors and heavy interior doors like glazed French doors or solid-core units.

Standard door casings are milled pine, available in the plain, gently curved shape known as clamshell and various colonial derivatives, which Apple Corps calls colonial clam. We prefer flat, square-edged casings. If the head (top) casing is milled somewhat thicker than those on the sides, so much the better. (I'll thrash over this subject more thoroughly later, in the section on trim.)

As you can see, there are many details you'll need to order doors, and many chances for mistakes. I imagine myself standing in front of each door and operating it when I make out my schedule. A door schedule has nothing to do with time but is a list of all the doors and their details. A schedule might have categories listed like this:

DOOR NO. Corresponds to numbers on the blueprints, to keep track of which door is which. Every door on the print should be numbered.

SIZE Width first, then height, then thickness (1⅜″ is standard interior, 1¾″ exterior)

MATERIAL Lauan, birch, pine, fir, Masonite, other

TYPE Hollow core, solid core, solid wood, pressed

HAND Left or right. Also used to describe pocket, bifold, French, etc.

FRAME Means jambs. Prehung, split jambs, solid jambs, rabbeted jambs, and how wide?

CASING If any, described, and applied or K.D. (knocked down), which means precut but not nailed on. Applied dictates split jambs, or one side only on solid jamb doors.

FIRE RATING If required by building codes

HARDWARE Lockset type number if there are several, and special hinge finish or type, and perhaps the number of the room (from the blueprints) that gets the lock operator

REMARKS Include glass, louvers, unusual finish, etc.

Each and every door in the building gets on this list, which is both the order for the supplier and the instructions for the installer. Care is necessary, as it's easy to overlook a small change on one door among many.

When it's time to install the doors, assuming they're prehung, I first check to see that the floor is level where the door will go. This is particularly important in remodeling. I like to leave one inch clear under a door in case someone puts a rug near it. If the floor is underlayment for carpet or vinyl, I figure that in. If the floor is wood, I sand the area where the jamb will rest with a belt sander, to clean it up and flatten it so the bottom of the jamb will fit better. If the level shows a high side, I calculate the clearance from there, and let it be bigger on the low side. If there is another door close by, I check the floor there, too, to ensure I can align the tops of both. The object of all this is to get the tops of the doors level, and aligned with their neighbors if necessary. Other doors a few feet away won't show a fraction of an inch difference in height.

Door jambs come long, like dress slacks, so the installer can cut them to the proper height. I cut the jambs to individual length to get the top of the jamb level and the door the right height off the floor. I set the unit in the rough opening and check the level and plumb (vertical level) of door and jamb, the evenness of the reveal (the thin gap between the door and the jamb), and the fit of the jambs to the floor if it's wood.

I then pull the nail that holds the door temporarily to the jamb on the knob side, and shim the unit into the rough opening. This means three or four sets of shims on the hinge side and two or three on the lock side. I nail through the jambs and shims into the framing, making sure everything stays put all the while. I like a 3⁄32-inch reveal all around, slightly less if it's humid weather and more if it's exceptionally dry. I check the swing of the door then, and can fuss with the hinges if things aren't exactly so. Doors take some time to install right.

When I hang the doors myself, I usually make up solid, rabbeted jambs. They are satisfyingly sturdy and alleviate the problem of loose stops. One drawback to this jamb style is that the stops can't be adjusted to conform to a warped door. Prehung or not, I make sure the jambs are the same width as the wall thickness they'll fit into, within about a sixteenth of an inch. Fudging is possible here by shaving the backs of the casing or even mashing the sheetrock behind the casing a bit, but each fudge takes away a little from a proper fit.

Pocket doors are a whole different kettle of fish. A pocket door, remember, is a door that slides into a recess, or pocket, in the wall. It's traditional use was between the front hall and the parlor, or between dining and living rooms, where it was often installed in pairs. Formal rooms in some houses are rarely used, and closing them off with a difficult to operate door is not a hardship. Nowadays, pockets are called on as a solution to a floor plan dilemma. They do work, but like sliding glass exterior doors, moving one calls on seldom-used muscles. If the hardware prevents the door from rolling smoothly, chances are the door will stay in its pocket most of its life.

Fifty to one hundred years ago in the heyday of pockets, the hardware looked like something for a barn door, with cast-iron wheels riding on I-beam tracks. The trucks (sets of wheels) were bulky and sometimes featured oil cups and threaded adjusters for servicing. The doors were often heavy and ornate, probably hard to get sliding even when everything was new. Wear in the hardware and settling of the house has made many of these old-timers unusable.

The modern version of pocket hardware has pushed the limits of frailty. The track is now roll-formed light steel, and the trucks are thin stamped steel brackets holding nylon wheels, one at each end of the door. Worse yet are the wall-framing substitutes sold with these hardware sets. The framing is light-gauge steel wrapped around thin billets of wood, barely enough to attach anything to. If you follow the instructions perfectly, the best you'll get is a flimsy wall enclosing a poorly hung door, bound to rattle and stick.

The best solution to this is having one custom built, in a thicker wall, practical if the rest of the wall is short. Sometimes the factory-made wall supports, overlaid with plywood then five-eighth-inch sheetrock, prove stiff enough. National's heavy-duty hardware can be made to work pretty well. Ideal, though, is eliminating pocket doors from the plan. The contraptions cost twice as much and work a tenth as well as their hinged cousins.

Another awkward door is the folding or accordion door. This is made up of wood panels about five inches wide hinged together to fold like pleats against the side of their opening. A track at the top of the opening holds the panels up on tiny rollers. Commercial models of this door are used as room dividers in schools and motel function rooms. They provide a way of closing an opening without a swinging door, but aren't intended for steady use. They carry a stiff price tag, too, so you really have to want one to specify one.

Reading about all these door types and styles and about door hardware and installation can't teach you how the door will feel when you use it. I strongly recommend that you tell your designer and builder you want good hardware, not the cheapest mass-market stuff. In your tour of prospective builders' completed work, try out a door or two and see for yourself how it works. If that's what you think of when you think of your doors, tell your builder. Getting your hands on the real thing is more informative than lots of words.

Trim

Trim, or architectural woodwork, means all the interior woodwork you see that isn't cabinetwork, but for our purposes I'll talk first about window and door casings, baseboards, paneling, and moldings. Trim is most often used to cover a joint between two different or separate materials or objects. It is also used to dress up a room, to make it grander or distinctive. And trim can be minimal or eliminated entirely.

You or your designer or architect will probably let the style of the house dictate the style of the trim. There is always a range of possibilities, though, so you do have some freedom of expression. As you can tell from looking at older houses, others in your position have chosen various solutions to trim design. Just make sure when you move from the outside of the house in you don't forget the concept.

One thing you'll likely learn about trim is that simpler isn't always cheaper. The extreme example is no trim at all, which forces the sheetrock or plaster crew and the finish carpentry crew to work together. The precise mating of disparate materials is tricky and time consuming, and labor costs outweigh any savings on trim material. Arranging a wall surface to bottom neatly on a wood floor, and then sand the floor, and finish both surfaces, is too fussy for most builders.

Likewise, you can buy precased doors and install them cheaper than you can hire the trim work done on the site. The factory's assembly doesn't provide any individuality nor does it cost much. The market-leading standard casing is thin milled pine, which is available in a few shapes and sizes. A simpler three- or four-inch flat casing costs more than colonial clam, and you may not even like it as well. The lesson is, deviation from the standard form costs you money, whether you go toward simpler or fancier results.

Here I'm going to lecture you about making all the trim in your house the same in style and finish, and dimension where possible. It's comforting to move from room to room and find similar woodwork, and I believe your house should be comforting. Comfort in this case means not having to think, when glancing at the woodwork, "Why is this different?" Even if you radically change the floor plan, your house will retain its original flavor. Inconsistencies show up as reduced value in the marketplace, too, as you may notice when looking at houses with a real estate agent. I don't care much if the inside closet trim isn't as fancy as the trim in the living room, though if you find a house in which it is, you know the original owner wasn't skimping.

Many, many remodeled or added-to houses get new trim styles with their new rooms. This is a way to save money, since older moldings are not easy to come by or duplicate. But I defy you to find an example of this work that in five years or so gives the satisfaction it should. I think a house and owner so treated have been badly used. If you think of your house as wholly yours and you want it to keep its value, you won't mix it up inside.

Think about the finished surfaces of a room. First there's the painted sheetrock of the walls and ceiling. You'll have a door or two, made out of wood, I hope. The finish floor is, say, red oak strip. Two walls contain a window each. Each of the elements in this so far rather plain room is installed in a separate operation, often by different individuals or crews. Nobody's going to get into anyone else's territory, so each installation stops a fraction short of the next guy's. The finish carpenter's job is to neaten up the place, to bridge the gaps between the different elements.

The first trim to install is the door casings. This uses up the longer pieces of stock first and is often simpler than windows — good for warming up. Casing the doors is easy as long as the jambs are flush with the wall surfaces. It's important that the casings fit tightly against the floor, unless it's to be carpeted. More important is the way the casings fit each other at the top corners.

Most door casings are mitered at the corners. That means the ends where boards join are cut at 45 degrees so they form a right angle. The 45-degree cut is necessary because the profiles of most standard casings are asymmetrical. Miters work well only when the wood is perfectly dry and the wall is perfectly flat. You

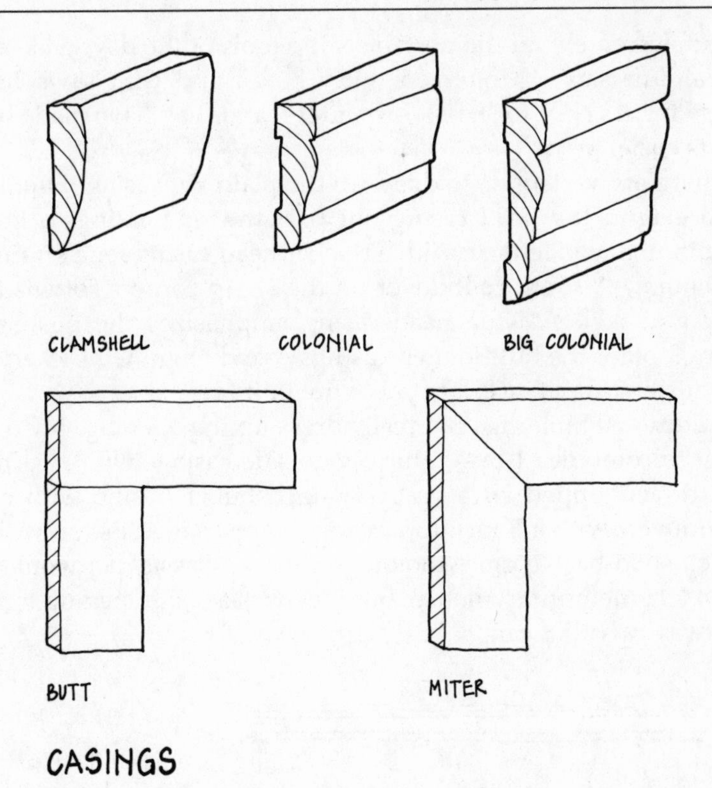

can make picture-frame miters fit because that's all they have to do. On a door, though, the casings lend support to the door frame, and must be secured to the walls and the jambs. Getting the casings tight to the wall sometimes sacrifices the miter fit, and vice versa.

The other trouble with mitered casing joints is their sensitivity to irregular shrinking and swelling of the lumber. If the moisture content of the casing changes very much in reaching equilibrium with house conditions, the joint will open up. This may happen to butt-jointed casings, too, but you can't tighten up a miter gone awry except by removing the board and recutting the joint. On a butt-jointed casing with its flat edges, you merely nail the top casing down a little tighter and your joint fits again.

A butt joint is simple. I bring the side casings up to slightly above the door, then set the top casing down onto their square-cut tops. I started out with butt joints because they were easier to cut accurately on the portable shop tools of the day, table and radial-arm saws. Motorized miter boxes and trim saws have made miters easy for finish carpenters, and they have made butt joints easier still.

There are variations to dress up the plain flat casing. Simplest is to extend the head casing out past the side casings a little. Combine that extension with a thicker head casing for a stronger statement. A slight roundover on the sharp corners softens the look and feel. A wide head casing emphasizes the post-and-beam look of the butt-jointed casing. Avoid too much trickery in the design, for chances are you'll tire of it fast.

Additional moldings can really dress up the opening. My own favorite combines 1-by-4½-inch-wide side casings with a ¼-inch roundover topped by a 7⁄16-by-1¼-inch bullet nosing with end roundover, with a 6-inch top casing wrapped near its top with a 2-inch solid-back cornice molding. This is a classic that wouldn't go in a contemporary house, but it evokes a solid opening to me in a way no other can.

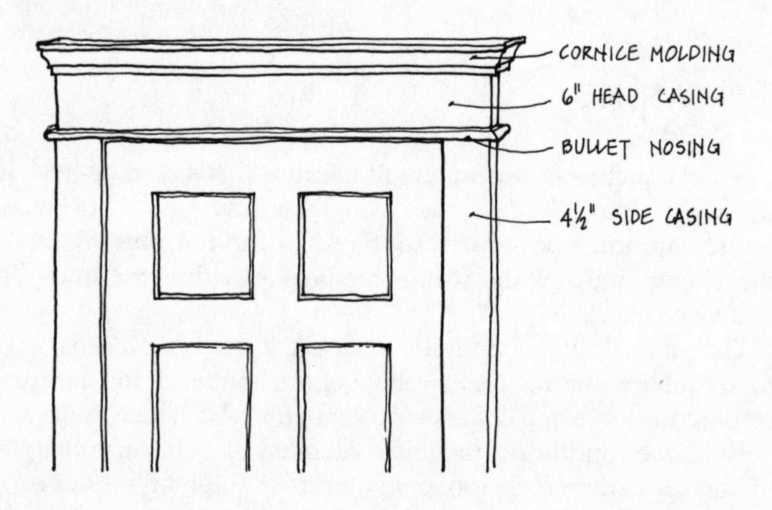

MY FAVORITE CASING

When you choose your door casings you are choosing your window casings at the same time. I see no reason to make door and window trim different in style, though a slight change in scale might be good if either doors or windows dominate the rooms. Windows, of course, have a real bottom to them, the sill, which doors don't, so they involve a few more decisions. If you're being faithful to the general style of your house's period, the sill will follow. I only urge you not to make the sill too narrow or eliminate it entirely from any window you can reach from the floor. The sill and the apron, the piece under the sill that finishes off the opening, should balance in visual weight the statement started by the head casing. Find one you like somewhere and copy it if you want, and be sure to figure out the components before the house design work is finished.

Baseboards range from the multipart Victorian model to the modern vinyl strip. Vinyl has little to recommend it but price, as it doesn't age well and never looks like much more than it is, the cheapest thing going. Stock baseboard is 3½-inch or 4½-inch colonial or clam base, similar to the casing stock, and 9⁄16 inch thick. This is a skimpy baseboard from most angles, and is out of place unless the trim style of the house is strictly conventional. I'd rather see a six- or eight-inch board, stating solidly that the base of the wall is *here*. Wide and thick baseboards fit many styles of house, and an additional molding on top can define things further.

In the plaster days, baseboards were installed before the finish floor, after the plaster was well dried. Any gap between the floorboards and the baseboard was covered with a base shoe, and the base was capped with a base molding. The shoe was nailed to the floor, not the baseboard, so that any settling of the floor wouldn't show. The base molding was nailed to the wall, so if the stiffer baseboard didn't conform quite to the wall, the molding compensated. This is a great system, but costly because of the many parts involved.

Nowadays the base is installed after the wood flooring. This takes time because the flooring when first laid is often not entirely flat, and the baseboard must be made to fit it tightly. Still, it's faster than fitting every floorboard perfectly at both ends. I usually run a belt sander over the flooring next to the wall to flatten out high spots before I try fitting the base.

The tricky part of installing baseboards is at the corners. If you're using stock baseboards with a molded profile, you must cope the inside corner joints, a fussy operation. (Coping is shaping the end of one board to match the edge of another it meets in a corner.) Outside corners take a lot of wear from vacuum cleaners and furniture, yet must fit tightly in a delicate position. Both kinds of corners are fit by a grown man kneeling and bent over and wishing he could straighten up and go nail on something more interesting. A nasty conspiracy neatly handled by wood and glue.

Inside corners must never be mitered together. Since the two baseboards must meet at 90 degrees, the miter would seem logical, but it just doesn't work. If you miter the joint and fit it meticulously, you're virtually guaranteed to open it up when you nail it to the wall. The trouble is, the nailing is working to push the two halves of the joint apart, and there's not enough solid stuff behind the corner to resist the push. An easier and superior joint is a plain butt joint. Once you fit the first piece in tightly, the second can only make the joint tighter, and you don't even have to nail the first near the corner to risk the joint.

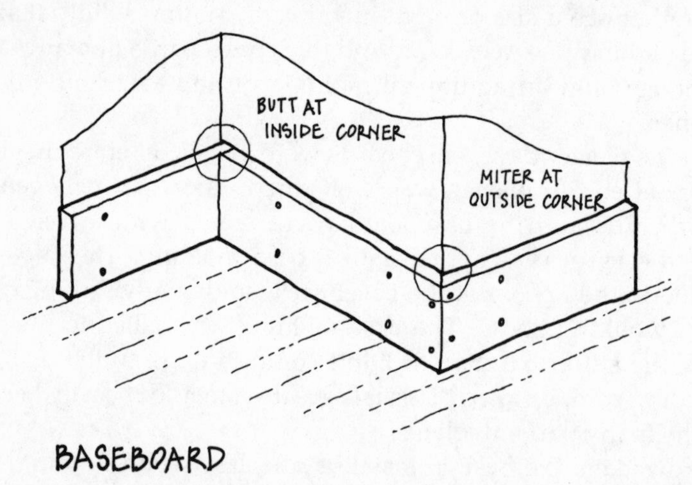

BASEBOARD

An outside corner in a sheetrocked wall is rarely exactly 90 degrees. If the miter — perfectly appropriate here — is to be sturdy, the two halves of the joint must make good contact with each other. So it takes fussing with most outside corners to find the right angle cuts. It's possible to cheat a little, to cut off the wood at the hidden side of the joint to make the visible edges tighter. Taking too much off means the pieces won't glue up well. (Every miter joint in every stick of trim should be glued and nailed.) Taking time to make these joints fit right rarely shows much at first, but will pay off in a sturdy job that will last decades.

Apple Corps' customers have lately reflected a renewed interest in more traditional house styles. Trim in these designs runs the gamut, from chair rails and crown moldings in the formal rooms to some paneling or wainscoting. Fireplaces are coming back, and some have elaborate surrounds of panels and stiles and rails. In most cases our customers are trying not to duplicate the designs of another time but to evoke them.

Chair rails and crown moldings are easy to install as long as you have good nailing in the walls. Crowns bridge the joint between wall and ceiling, and the top plate of the framed wall should be in the right place to accept nails, just at the top of the wall. If the ceiling is flat the crown is a breeze, providing the room doesn't have too many ins and outs. I like to set up plank staging so I can easily reach the top of the wall, and have someone cutting pieces that I measure and nail up. I make up all the midwall butt joints with 45-degree angles so a tiny adjustment is possible. I miter, glue, and nail the outside corners and cope the inside corners, and tack the pieces up until I'm sure everything fits, then nail the whole works at once.

Chair rails, meant to keep chairs from damaging walls when they're shoved back, should be fairly simple unless the rest of the trim is quite complicated. I find it very helpful to have nailed, at the height of the chair rail, two-by blocking in the wall before it's insulated. Using two-foot stud centers is great in the outside walls, but doesn't provide quite enough nailing for thin chair rails. If the room ceilings are a consistent height, I measure down to get the chair rail height, to avoid problems caused by discrepancies in floor heights from room to room.

The chair rail may be the topmost feature of wainscoting, which originally referred to paneling of any sort. We now accept it to mean paneling of approximately the lowest third of a wall, whether in wood or tile or even wallpaper. ("Wainscoting" comes from the Middle Dutch word for a wagon partition. The English word came to describe the superior grade of oak boards used in the wagons, thence paneling in general.) The specific design of the wainscoting will dictate the framing it requires. Scale drawings of the paneling are the only way all parties can know what to expect.

A wall can also be completely paneled. This can be done beautifully, but it's important that it be integrated into the design and finish of the rest of the room. Sometimes a light-toned wood like birch or maple satisfies the urge to have wood on the wall without the heavy feel of ever-darkening oak or cherry. And if the room can handle it, these darker boards can be really stunning. Care is once again the order of the day, especially in the design. With good nailing assured ahead of time, nailing boards on the wall is no big deal.

Archways, those doorless openings between rooms, should repeat the trim of the doors and windows. You may want to carry the same theme on smaller trim items like medicine cabinets and laundry chute doors. Be sure to reduce the scale of the trim accordingly or it will cease to be charming and become overbearing. Remember, the more you can keep the theme consistent throughout the house, the more it will be evident that you knew what you were doing. The builder helps as best he can by providing sturdy craftsmanship, but nothing beats a good design.

Stairs

In many houses these days the major item of interior finish work done on the site is the staircase. Stairs have many pieces and many close fits. Stairs and railings, along with kitchen cabinets and the back door, take a lot of pounding in the house and are expected to shine on through regardless. They are the place of many household accidents that proper construction could mitigate. Stairs cost plenty, which is one reason ranch houses are so popular with developers. And they can be very attractive, and worth the investment of time in their design and construction.

Stairs aren't always built on the site. In most parts of the country there are stair-building companies that will produce and install a factory-built staircase. If the company is a good one, the result is identical to a well-done site-built stair. Measuring and consulting are crucial, as is the skill of the installer, but that's true whoever builds it. If a conscientious builder recommends to you a custom stair subcontractor, take his advice. I like building stairs myself, and would be loath to farm the job out, but a fine product can be yours either way.

No matter who builds your stairs, the first part of the actual construction is building the rough stairs. This consists of figuring out the exact dimensions of the finished staircase, then building a rough version of it for the duration of the house construction. The dimensions really matter, and are set by the building code in most places. Codes, in their endless quest to quantify all aspects of building, often tend to mandate sameness in the name of order. You may lust for a radical-looking staircase, but you still have to be able to get up and down it safely every time.

If a staircase is totally enclosed in a hallway, it's called closed on both sides. (This is the standard for many commercial applications, since the stairs then can be isolated to contain a fire.)

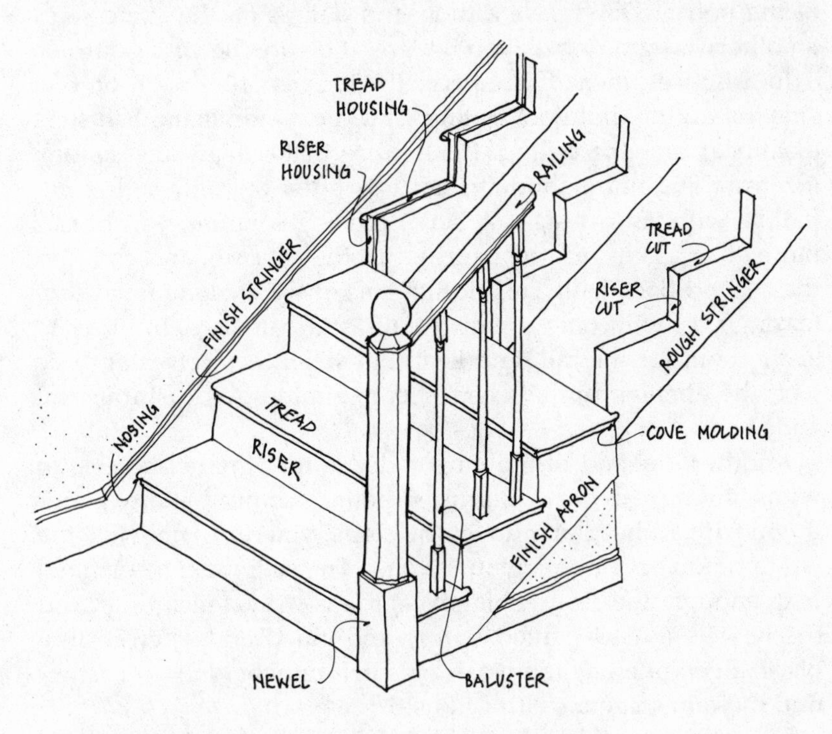

STAIR DETAILS

Many stairs have a balustrade (rail and pickets) on one side; that side is called open. In a larger hall, stairs might be open on both sides for at least part of the flight. A timber-framed house might suggest completely open stairs with just large-scale stringers and treads, called Alpine style by some. Grander still are curved stairs, extremely difficult to construct in the hurly-burly of a building site. Most construction techniques inhere to any staircase design, but I refer you to your builder and architect for the details.

The least expensive way out is the closed stair, which has no balusters, the delicate vertical pickets holding up the rail. Parts and labor for an open stair are more expensive, with finer details, especially the railings. Railings that spiral snaillike at the bottom over rounded treads look fancier and cost much more than the closet pole on brackets typical of a closed staircase. The main thing to be sure of is that you fit the style of the staircase to that of the house. Don't save a thousand dollars on the staircase in an otherwise grand hall; you'll regret it before the year's out.

In a house pinched for space, the stair is often open on one side just to the first-floor ceiling, and the handrail and balusters disappear into the ceiling. It's hard to make this staircase look like anything but a shortcut, but it is often the only style compatible with your design. A better-looking solution, which uses more space, is to run the banister up the stairwell and end it in the second-floor hall. This design beckons you along instead of teasing you with only a hint of rails and balusters. In order to leave room for the rail, and keep the stairs themselves at three feet, the opening in the second floor should be at least four feet wide.

Another demand of building codes is that the rail be able to withstand a push of two hundred pounds applied at any point. If you tripped headlong down the stairs, you'd certainly like the rail to hold you, but many just won't. The stair must be planned with enough newels (the big posts) to make the rail strong. And the newels must be fitted tightly and glued and screwed into place. Everyone has felt weak rails before; wouldn't you rather find them in someone else's house?

I use construction adhesive behind treads and risers when I screw and nail them on. A friend uses silicone caulk; he figures it will stay flexible longer. On the open side I miter and glue the front tread corner and screw end nosings on, then plug the screw counterbores. I stick glue blocks under the stairs to secure tread and riser together. I use nails and adhesive to fasten the rough stringers to the walls. Every railing and baluster and molding and skirt and trim joint is glued.

There exist hundreds of variations on every theme of stair design. Within each design is a choice of woods and finishes. The repetitive, purposeful pattern of almost any staircase is an excellent format for its builder's craftsmanship. Insist that your stair be at least as well done and amply budgeted as any feature of your house's finish work.

15

Kitchens and Baths

KITCHENS ARE THE subject of innumerable magazine articles, books, and advertising pamphlets. They are discussed solemnly by planners, rapturously by manufacturers, and volubly by anyone you ask. When you're daydreaming about your new house, the kitchen will undoubtedly be foremost in your mind. You may be struck with information overload. Again, starting from the beginning makes everything simpler.

Probably the first thing most people think of in their kitchen is the cabinets and counters, and where the appliances go. I suggest you start further back. Locate the kitchen first in the circulation pattern of your life and your floor plan. Traffic flow and access matter more than almost anything else. Next comes interaction with other rooms, like dining areas, entrances, and decks, and with the outdoors, source of the highest-quality light and air, and of a sense of place.

As with the overall design of the house, always remember that your dreams will influence your choices. When you say "den," do you picture cozy afternoons reading in the waning sunlight? Or is it just as likely the neighborhood five-year-olds' Combat Rangers meeting is there *now*? Can you work in the kitchen near a refrigerator subject to regular raiding sorties by ravenous adolescents? If you enjoy frequent visitors, will you design a kitchen they can sit in while you work or watch a pot? Imagine the actual and design for it, with a few possibilities left open. The way you and your family live should be the determining agent of your planning.

Once you establish all these parameters, you can design a kitchen to fit them. By "design" I don't mean you to take over your architect's job. I mean note the features and goals of the kitchen, just as you did for the house as a whole. It'll help to write down what you and your family want, not things but ideas, not solutions but desires. Don't write "microwave," write "quick meals." Figure out what is most important to all of you on the list, and what order your priorities fall in. Don't dream, be realistic. Spouses must often compromise here. If you say, "Just let *her* decide about the kitchen, I don't really care," nobody will believe you for a moment.

Next, go to your designer with your list. Your discussions with your household should have produced a list of united goals. Here your preconceptions about the brand of cabinets or flooring may impede the architect's work. You may begin to think he's ignoring your wishes if he chooses some other brand or concept. If you approach the room with an open mind, you'll likely get a more personalized design. And remember, don't design your kitchen for its resale value. The next guy is apt to tear it out no matter what you build.

When you get the plan back from the architect, see if it satisfies most of the goals you designed for it. Try not to get involved in the intricacies of the plan, but think in general terms. Look at the forest, not the trees. It's easy to be bowled over by a finished drawing, so prepare to resist.

You'll probably get the kitchen plan along with the entire set of house drawings. Give yourself plenty of time to mull over the plan as a whole, then in detail. If you need help thinking about your kitchen, check the design sections of Sam Clark's *The Motion-Minded Kitchen* (Houghton Mifflin, 1983).

Once you have a plan you approve of, you may have to pick out the things in it. This is something you can get loads of advice about, especially from salesmen. Before you choose flooring, counters, or cabinets for your kitchen, go see them. Your architect may submit a great plan to you but miss your taste by a mile. You'll just have to get involved. Since there are so many choices, I can't hope to address them all. Instead, I'll scatter some of my general opinions about kitchen products before you. You'll have to decide for yourself anyway.

I've already covered windows and doors in chapter 9. For floors, the choices are usually vinyl, tile, wood, or some combination thereof. I prefer few seams in a kitchen floor, whether between two different floorings or between pieces of the same stuff. So vinyl is good, though the patterns may not suit you. Most vinyl tries to make a strong statement of its own, and may clash with the other elements of the room. If you can find one you like, it makes a good floor, easy to care for and easy to replace when you tire of it.

Ceramic tile is another matter. Tile can really dress up a kitchen. Colors and styles are available to complement diverse kitchen designs and equipment. I caution you to avoid using too bold a pattern or texture or color. A room in which the surfaces' qualities enhance each other works better than one in which each feature strives for ascendency.

All tiles demand more maintenance than almost any other flooring. Grout gets loose or filled with dirt and is sometimes tough to clean. If you do use tile, specify that the installer use thinset mortar or high-quality mastic, like H. B. Fuller's Double Duty, to set it. And insist that the grout be mixed with a latex additive. Your tile floor will be expensive and last a long time, and you'd best be sure of a first-class installation.

Tile of any sort is hard, which means anything breakable breaks on impact. Vinyl and even wood are slightly more forgiving. Tile is hard to walk on, tires your feet, and is usually very conductive of thermal energy, so cool tiles will make your bare feet colder faster than almost any other flooring. (This is especially true, and problematical, in bathrooms.) Glazed tiles are slippery but that makes them easy to clean. Unglazed tiles like quarries are harder to clean because they're more absorbent, but they can get a little dirty without looking terrible. Quarries' wide joints are tough to clean, in the burger joint or your kitchen.

Wood floors survive in some kitchens and not in others. If

your yard or walk isn't paved and much dirt comes into your kitchen, stay away from wood. If you're a seal at the sink and the floor will often be wet, the dirt on the floor will become a grinding paste — death to the wood floor's finish. If your kitchen floor will stay quite clean and dry, and you like the warmth of a wood floor, have one by all means. I prefer the harder, tighter-grained woods for kitchens, oak being good and maple much better. Avoid a softwood floor — it's tough to stay ahead of the maintenance. In any case, four or five coats of good urethane isn't too much for a kitchen.

When it's time to buy cabinets you can easily get bewildered. Within the confines of the style you seek, I recommend you judge the cabinets by their hardware. Drawers should be supported by full-length runners on each side, not one down the bottom middle and little wheels at the front corners. Best are full-suspension slides that let you open the drawer fully out of the cabinet and that support it well for the whole distance. I like the models with a row of small ball bearings trapped in tracks; they slide like expensive file cabinet drawers, smooth and quiet. Few factory cabinets have these, though, as they cost about twenty dollars a pair instead of five.

You're more likely to see three-quarter suspension slides with nylon wheels. These are fine and will last a long time if they stay firmly attached to the drawer and cabinet. I've seen expensive cabinets come through with the drawer slides falling right off the cabinet frames. Since you are likely inexpert at judging hardware quality, your best approach is to look at several brands and price ranges. Looking at five examples will enable you to judge the lot.

Many of the cabinets produced today use complicated European hinges (or their domestic counterparts). These are an installer's dream, for they allow him to adjust the doors in three directions. They also make it possible for doors to open 180 degrees. They're great. I've also seen a couple of three-year-old cabinets with the same kind of hinges that were so worn out they couldn't be adjusted back into line. I still like them, and I also recommend simpler pin hinges where possible. Pin hinges show in the room, and work best with overlay doors, but they last a long time and cost little to replace.

Lazy Susans are something you either like or don't. I don't, but your kitchen is none of my business. Little soap trays made with the blind panel in front of most kitchen sinks are cute. Pull-out shelves work all right, I guess, though I think a simple shelf is the most flexible. Some cabinet lines come with twenty or thirty options, from spice racks in drawers to leaded glass doors. Get what you want and can afford; just remember, you have to use this stuff, too.

Cabinet features are not something I care much about, but I do care how the things look and work. I like simple cabinets with few moving parts and open shelves where practical. I prefer traditional-looking cabinets with panel doors, though I like flat panels better than raised. I'm tired of looking at Eurostyle cabinets with Melamine or laminate doors and no visible frames. But I care most that the cabinets work well and long.

Pantry cabinets are handy as long as they are strong and close to your work area. A desk in the kitchen is nice, though it may lend itself more to being a temporary repository of everyone's parcels than to the advertisements' penning of sweet notes to friends. A baking cabinet or area with a lower counter height is useful to some, and great for small kids when they're learning to cook. Lots of cabinet details can be customized if need be for someone who has physical difficulty with components of standard dimensions. Custom cabinets can solve all these problems.

Custom cabinets are generally slightly more expensive than the equivalent high-quality factory jobs, though they can't compete in price with the cheapies. You're able to specify which hardware you want, from drawer slides to doorknobs. You can get just the wood you want, and just the design. You can support the small industries of your area. You can get predictable delivery, though you might have to accept a long waiting time. You can get the cabinets made to fit your kitchen and your desires. Why go elsewhere?

Countertops embody many of the same decisions as kitchen floors, though there are a few new players. Tile and wood are popular, though the leader is plastic laminate, of which Formica is an example. Composite materials such as Corian are gaining ground. On the expensive side are natural marble, granite, and slate. Counters must deal with heat and cold, sharp objects, abrasive cleaners, dropped cast-iron frying pans, flooding, baking in the sun, and spilled grape juice. No one material survives perfectly all these trials.

Wood counters mustn't be cut on unless you are willing to refinish them constantly. Of course, laminate counters are worse, and tile will kill your knives. Laminate takes some heat and cold all right, but not pans right off the stove. Lots of wet scrubbing is murder on wood. Abrasives dull laminates but are fine on tile. Wood should never, *never* be used directly around a sink, period. Tile is also a problem around a sink, unless the tile is very flat and smooth so the sink can fit tight against it and keep water out of the sink-tile crack.

Most wood counters installed these days are butcher block, usually strips of rock maple glued together in counter top widths and available in several lengths. These are generally long lasting if the finish is maintained according to the manufacturer's instructions. Using one for a cutting surface limits you to refinishing with nontoxic, penetrating finishes such as salad oil. If you don't mind obvious signs of use, or won't cut on this stuff at all, butcher blocks are great. But not around the sink.

I've also used plank tops, particularly for peninsula or island counters not confined on two or three sides. A plank top is just thick boards edge-glued together to form a wide surface. The flat grain of the wood, thought by many to be the prettiest, is exposed. This kind of top moves a bit with the seasons, which is why I prefer it in an unfettered location. It requires maintenance, hot or rusty pan bottoms stain it, and knives cut it, but it looks great. Stick with the hardwoods for this unless you want a very rustic top.

Laminates work best over plywood or dense particleboard underlayments. Around the sink cut, the underlay is exposed to water if the sink leaks at the countertop, so I sometimes coat the cut with urethane or some silicone caulk. I try to buy the laminate, which comes in many sizes, in pieces that minimize the number of joints. Even distribution of the contact cement, applying strong, even pressure right after contact, and care in fitting joints make a lasting installation. The postformed laminate counters that lumber yards sell you by the running foot don't appeal to me aesthetically, though they are inexpensive and have no joint between counter and backsplash.

A laminate that is solid in color all the way through was introduced in the early eighties. (Most laminates are made of layers of resin-impregnated paper with a colored plastic coating bonded to the top at high pressure.) Color-Core is Formica's trade name for this product. It costs two or three times what regular laminate costs, but has the advantage that edge seams don't show the dark line of the older style. Though it's great for that, I find Color-Core to have a soft surface; I don't think it will stand the test of time.

DuPont's Corian and its ilk are fairly new to the market. They are made of a heavy and dense polyester resin that, unlike most counter materials, is homogeneous, the same stuff all the way through. If you damage it with a sharp knife, you can sand it smooth again. Corian is pricey and available in a limited but growing number of colors. It can be extensively customized by a skilled woodworker, and makes it into lots of fancy kitchens and baths.

Tile tops should be set on a double layer (or a firmly supported single layer) of three-quarter-inch fir plywood. I often seal the plywood with urethane, especially around sink openings. Ideally, a chlorinated polyethylene (CPE) tile membrane should go over the ply before setting the tile on it. In most cases a membrane isn't necessary, as counters just don't get soaked all that often. What is required, though, is a first-class grouting job, with the grout mixed with latex additive, and all the joints firmly packed and smoothly finished off.

Both laminate and tile counters must be finished on the front edge, where you can use the same or a contrasting material. I have often used wood facings with laminate and tile tops. It's critical to coat completely the edge of the particleboard with silicone caulk before screwing on the wood. This seals wood and underlayment against water invasion. Wood facings are pretty but slightly temperamental; they don't always work right in front of the sink.

I have made several kitchen counters out of used blackboard slate stuck to plywood, with a wood edge. The slate is about three eighths of an inch thick, rugged enough for the occasional hard bump. It can be cut with carbide-grit abrasive tools. It is generally pretty smooth, though I inspect each piece before I buy it, usually at a used building materials supplier. Slate makes a great-looking counter, unusual and long wearing, if a bit hard to clean.

At the top end of the scale are granite and marble. These should be polished and perhaps sealed — ask the supplier. Cabinets must be rugged enough to take their hundreds of pounds and hold them rigid. Expect some delay in getting your order filled. As with floor tiles, the stone counters are quite apt to be the coldest thing you'll touch in the room, and you might find that uncomfortable. And keep this stuff clean. You'll want to, of course, since it's so gorgeous.

Every counter top decision usually involves a backsplash choice. The backsplash is whatever goes against the wall behind the counter to make the back of the counter easy to clean. Simplest is a board, painted or varnished. It's all you need if the finish is sound, but should be well screwed to the back of the counter before the counter is installed. That way the joint between counter and backsplash can be truly tight.

Another way to go is a laminate covering over the board, which in this case is usually plywood. This choice is best with laminate counters, of course, and must be carefully glued to be long lasting. Some people, including me, find them a bit fussy. Postformed counters avoid the backsplash problem by bending the laminate over a curved base at the front, leveling off across the counter, then sweeping up the back and to the wall in an unbroken piece. These work OK, but as I said earlier, I don't like them.

I often make a backsplash by cementing a sheet of the laminate right on the wall behind the counter. I put this sheet on before the countertop, and run it up behind the upper cabinets if there are any. This way the wall has a very simple appearance and is easy to clean to boot. The counter must be firmly attached to the wall, though, and the wall framing should be thoroughly dry and done moving around. A two-by-six let flatways into the studs behind such a joint makes a risky thing surer. I caulk the joint with silicone just before I slide the countertop into place, and often after. Alternatively, a metal coping joint can be used if you don't mind looking at it.

If the backsplash material is at all flexible, I like to scribe it — that is, fit it — to the wall behind, since I want a tight joint between backsplash and wall. If the material is rigid, like a board or Corian, I will try hard to make the wall straight every chance I get. All this fussing results in a long-lasting installation, easy to clean and nice looking.

In general, cabinets that don't go to the ceiling leave dust-collecting shelves on their tops. Lots of exposed woodwork, like beams or paneled ceilings, in kitchens often means tough and frequent cleaning jobs. The more complicated the face of the cabinet, the harder it is to clean. Make sure the counter will hang over the doors below; one inch is good. This inch is sometimes considerably reduced in clean, modern designs. It looks great, but everything that falls or drips off the counter gets a chance at the drawers and doors.

One thing to be wary of in a kitchen design is the combining of too many different elements, especially in counters. Part of what is pleasing about cabinetwork is the repetitive quality of the design. Those rows of drawers and doors and handles and hinges are rhythmically satisfying, the way stair balusters or fence pickets are. Intrude on that order cautiously. Having three different counter surfaces may seem functionally necessary, but often blows the looks of the whole thing. Further, joints between disparate materials are always a source of difficulties.

Appliances are not my department, but putting them in sometimes is. The biggest problem people run into is the dishwasher not fitting under the counter. Most dishwashers need thirty-four inches or more of free height. With a 36-inch countertop height and a 1½-inch-thick counter, all's well and good — unless the finish floor hasn't been installed yet. Nailing down a ¾-inch-thick floor can get you in deep dishwasher trouble with few solutions, none good.

Garbage grinders are required by local ordinance in some areas and are popular in any case. Models vary, but all are either batch or continuous feed. Continuous feed means just that: turn it on with a switch and shove the stuff in. Batch-feed models have a switch in the drain plug–cover, and must be closed up for each use. This is safer, especially with kids around, and requires no separate switch. On the other hand, it's slower and not so handy.

Built-in ovens are popular, since they're easy to use. Each brand and model requires a different cabinet opening, though, so be sure to supply the model number to your builder in good time. A drop-in range looks more custom-fitted than a free-standing model, and is; trouble may ensue when it's time to replace yours. If you specify a range or cooktop with a built-in fan to eliminate the range hood, let the builder know as early as possible. Ductwork for these units may require changes in the framing if it wasn't planned for. Again, good planning prevents most installation problems.

How about that range hood, anyway? First of all, forget the ductless hood, which is supposed to recirculate air but trap the grease and odors of cooking. It won't. For ducted hoods, duct runs should be very short, should be made of metal with as few corners as possible, and should go down and out instead of up and out if there's a choice. In the duct cap outdoors, the flap may not seal very well, and a down-then-out duct minimizes the siphoning of heat out of the kitchen. Moisture collects in these ducts, too, so it's preferable to head the stuff toward the basement.

Refrigerators are always a pain. They're big and there's never a good place to put them. On many you can't get the drawers out unless the door has room to swing all the way open. Manufacturers make all different sizes. Make sure your cabinets are planned around a specific refrigerator. A typical model won't fit in a row of cabinets facing, say, an island three feet away. The fridge sticks out three or four inches from the cabinets next to it, and the door is maybe thirty-three inches wide. Bang, the door hits the island counter.

One solution is a pricey but marvelous unit made by Sub-Zero and one or two others. These are just the depth of standard cabinets and don't come with finished sides or front. Your cabinetmaker adds his matching panels and builds the whole thing into a wall or cabinet. Great if you can swing it.

Most other planning problems come in the corners between cabinet runs. Drawers and doors in cabinet corners have an amazing tendency to hit their adjacent cousins. A drawer might come out and strike the protruding handle of the range. Adjacent door handles might catch, and they certainly block each other. Pay attention to the corner details in the design stage. Choose handles wisely to avoid interference. Try to keep appliances out of reach of corner drawers, even when the drawer is being pulled all the way out for removal.

Where to put switches, outlets, and lights is a vexed question. Some people like all such technical stuff hidden, and will gladly reach a bit for an outlet to keep it that way. Others would rather have everything in plain sight, easy to use and obviously more than adequate. And the building codes must be accommodated regarding the number and spacing of outlets. Lighting is tricky in kitchens, with task and general lighting both required. As you can see, the electrical layout in your kitchen must be designed along with everything else.

Plan a minimum of two separate twenty-amp circuits just for the outlets, which are generally put halfway between counter and upper cabinets. If you want them out of sight, your electrician can install Wiremold outlets in a strip under the upper cabinets, at the back. Since you can't see these outlets when you're standing up, they're hard to use; you must bend down to find them and reach up under the cupboard. Plan to leave most countertop appliances plugged in, and of course you'll see the cords even though you can't see the outlets. Wiremold is hard to install neatly and is more costly than regular outlets, but it's more convenient in that the outlets are more or less continuous. With standard outlets, plan where your toaster and blender and everything will go ahead of time and your kitchen will be neater.

Lighting is another matter. You'll need light on the work you'll be doing and general light in the room, too. I like under-cabinet lights, now that Lightolier makes a neat one. These are great for bright counters, though too dim to light up the room much. Recessed ceiling floodlights can light counters with no cupboards over them, using twenty-five to forty watts per foot of counter. Dark cabinets and counters, wall and floor coverings, will absorb more candlepower than light ones.

For room lighting, many of our customers like the clean simplicity of recessed fixtures. With flood bulbs flush to the ceiling surface, these provide a broad light pattern. A wider spread is available from a surface-mounted fixture. Styles of these and many other fixtures are from either the dark ages or the future. Good luck deciding. Hanging fixtures give more choices but watch out for headroom. An experienced lighting salesman or electrician can give you good advice about how much light you need, though your own experience and taste are more accurate arbiters.

In the end, I simply must release you on your own recognizance for your kitchen plan. Many people stand ready and willing to give advice, mostly about details and sometimes about the details they know best. Start with traffic, location in the building, and relationship to the outdoors. Fill in with your particular work habits and decide which order of cabinets, counters, aisles, and appliances will make those tasks easiest. See and touch everything you can to help you decide.

Bathrooms

Bathrooms have changed drastically over the years. From their humble functional origins tucked away in odd corners, they have gotten bigger and fancier until some are now as big as bedrooms. People don't often go that far, but most want their bathrooms convenient, airy, spacious, and good looking. Savings on bathroom budget sheets come from using standard fixtures and materials. Exotic tiles and faucets can double the cost of materials in an ordinary bath. The calculation may include the premise that the bath is small, so changes to a small part of the house can't cost that much, and won't it be nice! I agree.

As with kitchens, the layout of the room and the location in the house count for a lot, though baths aren't used as frequently. Make the first-floor powder room or bath easy to get to, and to get past. It's nice if the family bathroom is central, and essential that it not be reached only through a bedroom. Parents love their own bathroom, especially as the kids get past ten years old. Privacy matters in the room, accessibility in the house.

I suggest you follow the trends and make your bathrooms large enough so they feel like real rooms rather than closets. Any plan will work, though, that allows you to move around freely in the room. Specify moisture-resistant (MR) sheetrock for any room with a shower. If you think a sick or elderly or wheelchair-bound person will use the room much, plan grab rails and accessible outlets. And have the door swing out; if someone falls in the bath and is stuck in front of an inswing door, you can't get in to help. Don't use a lock on the door if you're afraid someone might use it who shouldn't. Paint the walls and ceilings with semigloss paint, and never texture the ceiling. Baths are best if not overstuffed; you want them to dry out between uses.

Try not to jam the toilet into a deep recess or corner or against a vanity. Strive for a minimum of 1½ feet of space on either side of the toilet's center line; 2 is better. This rule holds when the toilet is next to the long side of a tub, and makes this layout less of a compromise. It's nice if the paper holder has a place to go next to the toilet and not on the back wall. Most people like the paper holder roughly even with the front of the toilet and just above 2 feet off the floor. If you're going to have a bidet, plan to double the space allotted to the toilet for the pair of fixtures.

A vanity or sink is easier to use if it's not jammed against a wall on either side. Unless it's in a make-do powder room, the sink ought to sit in a minimum thirty-inch-wide space, more if it's recessed. Some people like their bathroom sinks higher than the standard thirty-two inches, and thirty-six-inch vanities are finally being introduced by the cabinet companies. If you want a recessed medicine cabinet over the sink, plan the framing for the exact one you'll use. Keep it eight or ten inches above the top of the sink.

You'll probably fool around with the fixture layout for quite a while and come away with a good plan. There just aren't that many choices and, given enough room, most will work. The smaller the room, the simpler should be the detailing; more space gives freedom in many ways. Just make sure you can stand and move anywhere you want to stand and move, and avoid designing a bathroom with sharp corners.

One problem comes when the bath is mostly against outside walls. In colder climates, pipes should not be installed in outside walls, where they are subject to freezing and themselves reduce the amount of insulation in the wall. I would never design a bathroom with the shower control valve in an outside wall. A vanity there could be piped from the side, around the corner in an inside wall. A toilet supply pipe, low down in a first-floor wall, gets little heat and can freeze. If yours is a cold region, ask your plumber how to avoid freezing pipes. Careful insulating is critical.

Assuming that you plan a room big enough to move around in, see if you can give it a window. I suggest moving it if you can't. A bath without a window must be ventilated mechanically to comply with most codes. A vent fan may work, but is best as an adjunct to, not a substitute for, the light and air from a window. Almost any modern window will work in a bathroom, provided you maintain the finish in good condition. (Refer to chapter 9 for ventilation guidelines.) A window isn't strictly necessary in a powder room. Remember not to put a window in the shower.

Artificial light makes lots of difference in this small room. A good light for shaving or making up is spread out and strong; both sides of the mirror should get fixtures. You might want a general room light, and perhaps one in the shower. Make these surface-mounted, especially in baths under insulated ceilings. Recessed fixtures channel warm air and its destructive moisture up into the cold attic, with a good chance of damage from condensation. As a variation, recessed lights can be installed in a lowered section of the ceiling called a soffit, sidestepping the insulation problem. Fan-light combinations and recessed heat lamps also must be well insulated from a cold space above.

Many building codes require ground-fault-interrupt (GFI) circuits for bathrooms and outdoors. In an electric circuit, the ground is the safe wire, one that never gets current and that is connected to, well, the ground. The third, rounded hole in a three-prong outlet is for the ground. The GFI circuit senses a false ground in the circuit — your wet hand, for instance — and immediately kills the power to the switches and outlets on that circuit when one occurs. Since electricity is most easily conducted by wet appliances and wet bodies, GFI circuits go in wet areas. Even if your code doesn't demand them, you should. The main objection to them, that they trip and kill the power all too easily, is one you can live with. Most of these devices are worth the very occasional aggravation.

As I said, ventilating fans are required in bathrooms without windows, mainly to remove moisture from bathing. Dumping that warm moist air into an outside soffit, or worse, into the attic, guarantees future problems with moisture and mold. It's best to mount the fan so the ductwork can go down an inside wall and exit the house just above the foundation. In baths with insulated ceilings, mounting the fan high in the wall works better than a ceiling installation because it keeps the insulation layer complete.

The worst manifestations of excess humidity in baths come in those with skylights. Codes allow using a skylight for ventilating a bath, but no one will willingly open one on a cold winter morning. Since the warm moist air rises, it will invariably condense on the skylight, even a triple-glazed one. This is a surefire dripper, and its trim can rot out in only a few years. A vent fan high in the room or in the skylight well, used assiduously, will prevent the damage. Again, exhaust this fan downward and insulate the ductwork if it runs through a cold attic.

Your house's overall heating system may provide the capacity to heat the bathroom but not give room-by-room control. Many people like a bathroom quite a bit warmer than the rest of the house, especially for showering. For them a heat lamp or two in the ceiling provides the extra warmth without too much cost. An electric baseboard heater or radiant ceiling heat can also do the job. Avoid electric fan-forced wall heaters, as they are noisy and have a questionable service life. A baseboard heater below the towel racks is great for warming towels. Keep heating baseboards well away from toilets; they won't last there, and someone might get a burn.

Houses with warm-air heating systems should have a good-sized supply register in the bath, but no return ducting. A return would distribute moisture and odors into the heating system. Since there is no air return, the bath door should stay open as much as possible to keep the room warm. And a door cut at least an inch off the finish floor will allow some air to circulate with the door shut. The heating system can't supply air to a room that is sealed up tight, any more than you can put your lips over the neck of a bottle and blow air into it.

If your house employs hot-water baseboard heat, consider using a modern radiant wall unit. These are wide-area radiators, flat and thin, maybe thirty inches square, and are installed from the floor up, against a wall. Some even have towel bars on top, granting the luxury of warm towels. New to the market and to me, these look like a good solution to the problem of getting enough heat into a bathroom, and they're attractive to boot.

Bath fixtures run from plain Jane to eleganto deluxe. The standard toilet, sink, and tub are the most widely sold and the cheapest the plumbing supply house has to offer. Even lower prices are available from discount building centers and home supply outlets, though often at the expense of reparability. It's probably true that the plumber who installs your fixtures and fittings is the one who will have to fix them. He won't like putting in something that's hard to repair.

The simplest plan for making a decision about fixtures is choosing one brand and sticking with it throughout the house. Avoid a name you've never heard of. With world trade bringing the goods of many nations to our shores the buyer has more options than ever before. But consider this: the main thing you want from your bathroom fixtures is long, dependable service. In this country, reliability is the consequence of not only good design and manufacture but wide acceptance by tradesmen. A jazzy Italian toilet won't do you much good if it won't flush and no one will fix it.

I think American manufacturers have tended to overbuild their durable consumer goods, and that has meant reliable products for the most part. European designs are modern and thorough, but I sometimes think the sales literature is of higher quality than the products. Quality means solid porcelain and brass, thoughtful design, and careful production, not just flashy features. Though plastic is everywhere and most products are getting more sophisticated and fragile, I still think building products of U.S. industry are pretty good. Plumbing fixtures are no exception.

Apple Corps built a new house a few years ago with state-of-the-art fixtures and fittings. Almost everything was imported from Europe, and it certainly looked great. Well, maybe the faucets were hard to work and the drains leaked a little. And the toilet didn't fit standard U.S. connections and had to be modified. There was some trouble because the shower valve wasn't the scald-preventing pressure-balancing type the code required, and *that* had to be changed. But didn't everything look wonderful? Wasn't everyone but the installing plumber happy? Why was he so frustrated, anyway?

Everything still looks OK, though the finish is starting to wear thin. The seals are leaking a bit here and there, and the lavatory drain must be adjusted frequently to work properly. When the kitchen faucet started to leak a while ago, it took eleven weeks just to get the parts. It's hard to find a plumber who will work on the stuff. The people still like the way it all looks, but they have yet to convince themselves they made the right purchases.

I must admit all is not perfect on this side of the ocean either. Some fixtures work better than others. I find that plumbers like to install the least expensive toilets, sinks, and tubs they think will last. Plumbers who service apartment or condo complexes, especially, seek the best combination of reliability and price. And different plumbers like different brands, sometimes having more to do with which supplier they use than with which brand is better — sometimes depending simply on what they're used to. You'll have to do your own research on fixtures to get what you want.

Your architect will likely specify the fixtures for you. Go to a plumbing supply house and see and touch the fixtures and fittings (faucets, controls, and so on) he recommends. If his choices are so exotic no local supplier keeps any in stock, you'll have to decide if the style and distinctiveness are worth the possibility of repair problems. Ask your plumber, or if possible your builder's plumber, what he thinks of the architect's choices. Selecting someone to ask may be tough, I know, since I still want you to get all this decided before you sign any agreement to have the house built. If you don't have a cooperative plumber, visiting two or three plumbing supply retailers should give you a sense of the options.

Some items simply must be decided beforehand, like the tub or shower. Your plan may include leeway in installing various vanities or toilets, but the framing for the shower or tub must exactly fit the unit you choose. Don't take a chance on slowing down and irritating your builder by waffling on your shower choice right up to when he's ready to frame for it. Even deciding on the smaller fixtures may carry implications for the framing crew, so don't wait until the hammers are poised to figure out what you want. Providing timely answers to your builder is crucial to your good relations with him, and keep in mind that fiberglass units, for instance, can sometimes take months to get.

Consider the water-saving aspects of current fixtures. Fresh water is scarce in some places, and sure to become more so. Toilets account for nearly half of everyone's water use, and modern units are designed, and mandated by code in some places, to save water. Where older models used four or five gallons per flush, now three is normal and less is possible. Plumbers take a dim view of this development because early water savers didn't flush well. Because of that, I think you should stay in the 2½- to 3-gallon range, unless someone can prove to you the reliability of a thriftier model.

Faucets and showers also now come with restricting washers to cut water flow. Use these or not as your conscience and predilections dictate. In showers I strongly recommend pressure-balancing valves. These ensure that the water temperature stays even, where you set it, no matter if the pressure fluctuates in the rest of the plumbing system. The valves were developed to eliminate the scaldings people got when they were showering and someone flushed a toilet or started the clothes washer. They work, and they're worth it.

I recommend china for your bathroom sink. The other choices are enameled steel and enameled cast iron. Steel is light and thin, and flexes enough eventually to chip off the enameling. Cast iron is rigid and the enamel lasts a long time, but many of these sinks seem to rust around the edges anyway. China sinks are durable, easy to clean, and inexpensive. The worst thing about them is that they're sometimes not flat on the lip where they sit on the counter.

Lavatories, as plumbers call bathroom sinks, can be installed under, flush with, or over the counter. The under-hung sink is a rarity, as it can be used only with countertops that are of homogeneous materials like marble or Corian. Rimmed sinks, installed by means of a stainless steel rim that rests on the countertop, won't work well on irregular surfaces like tile, and present two joints subject to standing water. Rimless sinks, china and some cast-iron models, sit on the counter and stick up above the counter's surface a little. I think these are the most waterproof installations, though some people prefer the sink to be completely flush to the countertop.

The neatest solution to the small vanity top is a one-piece counter-sink combination. This fella can't leak. It comes in twenty-five- or thirty-one-inch widths in china. You can also get a wider top, or a two-bowl top, in Corian or cultured marble or cultured onyx. The cultured products have improved vastly in the last few years and aren't subject to the crazing and cracking of those of even eight years ago. If you can find a vanity top–sink combo you like, use it.

If not, don't use wood, just as you wouldn't use butcher block around a kitchen sink. Plastic laminates are fine, very serviceable if somewhat boring. Tile means money and maintenance, but is flashy and enduring. Granite and marble are classy and solid. With any top, the detailing of the backsplash-counter and the sink-counter joints is important. Lots of water gets splashed around some bathroom sinks; be sure your design keeps it where it belongs.

Showers and tubs are simple fixtures but often problematical; they leak. If the leak is in a second-floor shower, the damage can be expensive. Most leaks come from the joint between the tub (or shower base) and the wall surface. Correct detailing of, or eliminating, the joint will usually keep things dry for decades.

You can eliminate the joint by using a one-piece shower or tub. These units, of fiberglass or acrylic, have swept the industry in the last fifteen years. Builders love them because customers ask for them, because they don't leak, and because they are relatively light and simple to install. The builder need not hire a tile setter, one more subcontractor to coordinate. Callbacks are minimized, since maintenance is restricted to cleaning. The units are fragile during construction, but hardy amid the rigors of daily life. Alas, they are too bland for the tastes of some, and most are too large to be used in remodeling projects. So tubs, and life, get more interesting with variety.

The old standard cast-iron tub with tile surround, neglected when fiberglass units became established, is now making a big comeback. Advances in tile-setting techniques and materials have made a nettlesome task simpler and relatively cheaper. The renewed interest has spurred diversity in the tile industry, so many choices from home and abroad are available. Tile is permanent, and has a strong presence. Choose your tile and your fixtures carefully, as both will endure many changes of the other colors and textures of the room.

Think of a tile shower job as one of two varieties. The first is tile walls atop a factory-made base, often a bathtub. The other is a shower that is completely tiled, including walls and base. The latter is quite different to build, and costs much more. In odd-shaped locations it's the only option available, but let's look at the factory job first.

In a shower or tub, water runs down the walls but sits for a while in the tub or shower base. The walls must shed water, the base must hold water — two different qualities. Using a factory-made base means the tile-setter and builder must plan only the walls. The critical part of this installation is the joint between the tile and the base, and details of this joint vary with different bases.

A common enameled cast-iron tub has a low lip around its outside rim, the part that's hidden in the walls. This lip isn't high enough to stop splashing water, but it will keep some water from running over it. The lip is really a secondary line of defense, something to take over if cracks appear in the visible joint. The tile job should be waterproof, right down to and including the tub joint.

The tub is usually the first finished item to appear in the interior of your new house. That's because the wallboard or tile backer, however they're finished, will sit on top of the tub or base. It's tough to protect the shining finish of a tub, and especially that of a shower base, but they must go in early, before sheetrock. We generally apply duct tape to the top surfaces of the tub, and fill the tub itself with batts of insulation. We're very careful, especially with fiberglass or acrylic tubs, to keep all metal out; sheetrock screws have an amazing affinity for tubs. Once the sheetrock or tile backer is in place but before it is taped, we often cover the tub with a piece of three-quarter-inch plywood. The sheetrock guys can stand on this for finishing, and the tub is safe from scratches.

The tub is set in place against the wall studding, with about a quarter inch of clearance all around. For tiled walls, we next nail on a fiberglass-reinforced mortar board, called Wonderboard. This product stays intact even when soaking wet, and no other common tile backer will do that. Much tile has been laid over sheetrock, moisture-resistant sheetrock, and plywood. MR sheetrock works until there's a leak, plywood will tolerate a minor leak, but Wonderboard works no matter what. The tile job is expensive enough to make the best solution obligatory, not optional.

I start out by framing the opening accurately. I add nailers at the height of the top of the tub, framing material to stiffen the wall in that area. After any insulation and vapor barriers I install a waterproof membrane on the studding, covering the first three feet or so above the tub. I lap the membrane in the corners and cement it, and let the bottom hang down over the tub an inch or so. I try to form the membrane snugly around the shower valve, to catch any leaks in that area. Later, when I set the Wonderboard, I'll trim the bottom of the membrane so it covers the tub's lip and extends out to the face of the Wonderboard.

I set the Wonderboard one-quarter inch above the tub, and run it up to just below where the tile is to stop, finishing the height of the wall with moisture-resistant sheetrock. (The tile then covers all but the smooth wall, and bears no hint of the work behind it.) I use fiberglass tape and thinset mortar to fill the joints in the Wonderboard. When I cut a hole in the backer for a soap dish or grab bar, I am careful to leave the membrane intact, to contain any leaks that might develop. When the thinset has dried, overnight, the backer is ready for tiling.

The layout of the tiles is part of the design work. Small tiles don't show up as distinct units but parts of a pattern. When the tiles are six by six or larger, they themselves become prominent. You may like some of your tiles laid diagonally; that's more expensive, so specify it ahead of time. If you want accent stripes of tile, or borders of contrasting units, get that clear before the contract stage. It's easy to double the price of materials for a tile job by just adding a fancy border. It's pretty hard to predict your color scheme without seeing the room, but at least you can decide on the general makeup of the tile installation. Tile jobs are perfect candidates for an allowance in the contract; you can choose the tile and pattern while you're standing in the room, and the contractor won't take a bath on the price.

I let the tile set for twenty-four to thirty-six hours before I grout it. Grouting is a routine job on glazed bath tile, a bit harder on ceramic mosaics (small sizes and varying glazes), and difficult on quarries and other unglazed tiles. Unglazed or matte-finish tiles aren't ideal in a bath, as they're tough to clean when the soap film builds up. I lay out the tiles to fit close to each other. It's best to make the grout lines as small as possible; the less grout showing, the fewer problems. I like to grout the vertical corners and the tub-tile joint with matching silicone caulk, rated for mildew resistance. Silicone can take a tiny movement in stride, while grout cracks.

Your tile job can cap a tub or shower base — of plastic, fiberglass, tile, or enameled cast iron — or simulated terrazzo. I recommend cast-iron tubs because I regard the installation as permanent, and they have the best track record. Fiberglass tubs warm up faster and come in more sizes and shapes and colors. Fiberglass or plastic shower bases are inexpensive and serviceable. Real terrazzo is made on the site by placing marble or granite chips in wet mortar, then grinding and polishing the surface after it dries. A simulated terrazzo base, molded at the factory, is heavy, fragile, and weak, with a lip too low to be splashproof. It is often poorly cast so it won't sit flat on the subfloor. Forget it.

A tiled base can be made leakproof, and the best way is with a membrane. A copper pan, soldered at seams and drain, is the old standard, and works fine if it isn't disturbed — for instance, by nearby remodeling. Old copper and solder lose their flexibility and membranes don't, theoretically. Membranes haven't been around long enough yet to be proven over time, but good producers generally use good research, and don't want to promote an inferior product. Drain fittings designed to work with membranes are commonly available, and make that joint safe. You can't oversee every installation in your house, but specifying that the membrane system be used according to the manufacturer's instructions is a good guide for the builder.

If you plan to use a shower curtain, the step-over lip of any shower base should be at least 4 inches high. The lip for a shower door can be lower, say 2½ inches. A chunk of marble makes a good top for this step, because its smooth surface means the door won't sit on irregular tiles, with the chance of a leak. Marble is not too expensive, usually less than it costs to tile the same area.

You'll have to plan the bathroom as a whole. Give yourself enough room to make it work. A foot lost from a bedroom is less noticeable than the same foot gained in the bath. Make the room easy to get to, and easy to get in and out of. Make it easy to maintain, with smooth surfaces and few little corners and crannies. Spend for halfway-tiled walls if you want, because the total expense is a small part of the house's price, and the result will return satisfaction out of proportion to the cost. Just don't spend too much on flashy tile and fixtures that fashion may disdain in a few years.

16

Painting

Exterior painting

Painting your house is something you may choose to do yourself. Most people have painted something by the time they can afford a house. By then they either find that painting is one of their talents or swear they will never do it again. If you're thinking of painting, you should calculate the time required to paint your whole house, at your speed, and consider the time you have available. If a three-coat exterior paint job on a new house takes three professionals a week, or 120 hours, how many two- or three-hour stints is that for you? Two months, perhaps three, of *all* your spare time? And don't forget to budget for the paint!

You certainly can do a good job painting if you care to. The pro makes his speed from a few shortcuts, but mostly from his ability to set up the job, put in a long, steady day with few interruptions, and keep his help moving around to his best advantage. You can do a better job than a careless professional, or maybe even a conscientious one, if you know how and take your time. But beware of diving into a big painting project without knowing what you are doing. Painting is hard and dirty and dangerous, so you decide.

I have already recommended that you paint your new house rather than stain it, and here's how. I'm assuming a well-built building, with clear pine trim and vertical-grain red cedar clapboards. The first step is to backprime the trim and siding. Backpriming is simply painting the back of the boards — the side against the house — with oil-base primer before they are installed. This cuts down on moisture penetrating the boards from back to front, and greatly extends the life of the lumber. It may cost an extra six or eight hundred dollars, but will save that amount on the first repainting and every one thereafter. And your siding will always look good.

When the trim is installed, prime its exposed surfaces, again with oil- or alkyd-based primer. Leave the cedar siding, though. Planed cedar clapboards come with a very hard surface, probably the result of machining and handling. Six months out in the weather will open up the surface grain so it will accept enough finish to last years. Too, pockets of tannic acid will bleed out at first, not after the finish is applied. After six months, or twelve or even eighteen, wash the siding with a fifty-fifty mixture of household bleach and water. The bleach opens the siding grain further.

When the siding is dried out again, paint the surface with oil-base primer. Primer needs a long time to fully dry, often forty-eight low-moisture hours. After that, brush on two coats of oil-base house paint, perhaps consulting consumer magazine tests to choose a brand. Completely avoid the house brands of discount chains. Gloss paint is a little harder than eggshell finish, but the eggshell is more forgiving, and cleans just as well.

With most natural finishes on siding it's wise to paint the trim before siding the house. An exception is when the finish product, like Benjamin Moore's Clear Wood Finish, must be sprayed on to ensure adequate penetration. In that case, paint on the trim will be wrecked by the sprayed material and have to be redone anyway. Galvanized nails can discolor clear-finished siding, so use stainless steel instead. With shingles, it's practical to dip them into a pail of the finish product before you nail them up. If the shingles are dry, you get good penetration and a lasting job this way, at the expense of using more finish.

Paint manufacturers are recommending that you use latex-base primer and paint on any house. Don't. Oil flows out better and levels, leaving a smooth surface. Over the years oil oxidizes, or chalks, so that at repainting time the new coat doesn't add much to the total film thickness. Latex builds up with each repainting, and the thicker film is less stable and more apt to flake or chip off. If the old finish deteriorates enough to expose bare wood, the oil product keeps it from drying out too much. The worst situation comes when a careless painter puts latex over dirty, poorly prepared oil. This finish is guaranteed to fall off, maybe in three years or less.

Preparation time costs the same as painting time, as far as the painter's concerned, and it's hard to get people who will do it. Maintaining your house's finish, therefore, is well worth your while. Because oil paint doesn't build up thickness, and its oxidized surface accepts recoating well, it pays to keep up with it. About every five years, wash the house and put another finish coat on all the oxidized surfaces. Different sides of the house react differently to the weather, so you may not need to recoat the shady side every time the sunny side needs it. Just don't wait until the finish has started to crack and peel.

Your choice of colors also will influence the paint's life. Dark colors absorb (and transmit) more heat. Sunny-side clapboards suffer more extremes of heat and cold in winter and over the seasons. The heat bleeds the natural resins out of the lumber, drying it out and promoting surface cracks. Dark-colored houses with clapboards over felt (tar paper) haven't a chance, painted or stained. I've seen houses barely thirty years old with the clapboards broken and falling into the yard, from weathering and poor maintenance.

Many folks now choose a stain finish so they'll never have to scrape old paint off. Though scraping can be minimized by maintaining your paint well, stain is ever more popular. The mistake with stain is thinking that since the house still looks more or less the same color as always, the siding is protected. Stain, since it leaves little film on the wood's surface, protects the wood less from moisture and the killer sun. It's critical, therefore, that the wood receive regular doses of restorative coats of stain, often every three years on the sunny side.

The house should remain unstained for as long as it would for

paint job, to open up the grain. Pressure-treated lumber
weather at least a year. The first stain coat should sit for
month or so before the second coat goes on. Most manufacturers recommend this, but it's seldom demanded by hurried builders. It's tricky to get an even coat of stain on the whole building. Lap marks, where one section dried before its neighbor's brush strokes blended with it, are a sign of a poorly planned job. Different shades on different sides are further evidence that the painter lacked experience or a demanding mandate.

There are some pitfalls to a good finish that are not under the painter's control. The general contractor must not rush the painter or insist he work under adverse circumstances. If you build the house properly from seasoned framing lumber and siding, control moisture migration through the walls, and observe due caution about the weather during the finishing work, you've done enough. Most else beyond that is the domain of the painter, and his responsibility as well.

Interior painting

The kindest thing you can do for your painter is give him enough time at the end of the job to do the whole interior, even if it delays moving day. You'll have a smoother paint job if the carpenters aren't making sawdust at the same time. Painting is skilled work, just like cabinetmaking. Though any paint job will transform your house, a good one will show it off.

Interior paint jobs can be divided into stain or paint "packages." The package refers to the way the trim will be finished — painted or stained. For a painted interior, the builder should finish all surfaces and install everything except electrical plates, appliances, bath accessories, toilets, and carpeting. The wood floors should be sanded and varnished, and the hardware installed on doors and windows. Then — and this is the ideal way to get a good finish — everyone goes away and leaves the painter alone to clean up once and go to it.

He should start by priming the walls and ceilings, caulking where necessary around the trim, and filling the nail holes. Next he will bring the trim to finish, with two finish coats of oil paint over a coat of oil primer, with everything carefully sanded and dusted between coats. The baseboards and window sills should be left a coat short of finished and be masked off, since they might receive a little splatter from the wall and ceiling rollers. Two coats of paint — latex is fine — are applied to walls and ceilings, with the walls finished last. The wall paint is then cut in to the trim around windows and doors, moldings and built-ins. ("Cutting in" means carefully painting right up to a surface that has a different finish.) Finally, baseboards and window sills are uncovered and painted their final coat. This is an efficient system, and it works best when the painter has the run of the house.

For a stain package, the painter's approach is different. First he'll stain all the trim before it's installed, apply a sanding sealer to raise the grain, sand it all, and apply a coat of varnish. Next he will prime the walls and ceilings, finish the ceilings, and put one finish coat on the walls. Then he'll go away, and the finish carpenters will install the trim. When that's done, the painter will return to finish the trim, filling nail holes and applying one or two more coats of varnish. He'll protect the sills and baseboards as for paint, and paint the walls and cut in to the trim.

There is some debate over using urethane varnish on trim. Urethane produces a harder finish and resists alcohol staining. As such, it is good on tables. On trim, though, urethane often seems to have a yellowish cast. It's harder than varnish, sure, but varnished trim lasts fifty years or more with little maintenance, so who cares? And varnish tolerates the smaller dings and dents of family life with less complaint. Too, for second or third coats, urethane is supposed to be reapplied within twenty-four hours, and that is often impossible on a big project.

If you're the owner of an older house, your painting options may have been exercised by previous tenants. A common mistake was to paint, with little preparation, over a varnish or shellac finish on doors and woodwork. The last guy's paint won't stick to the varnish underneath and your paint is just stuck to his. This combination gets shabby quickly every time, and can be cured only by heroic removal of the offending layers. A common old-house problem in New England is the calcimine ceiling. Calcimine is closer to whitewash than paint, and won't grab on to new paint enough to hold it in place. Calcimine removers rarely work, and the best solution is a new sheetrock ceiling or painstaking scraping and washing of the old ceiling, impossible if the plaster is weak. About all you can do with an old house with poor surfaces is get expert and patient help, and adjust your expectations.

Wallpaper ain't what it used to be. There are so many choices of finish, backing, color, pattern, material, adhesive, and price it's impossible to know what to get. There are a few rules, though. Don't buy your paper from a discount chain, or on sale at a regular source. Choose a reliable dealer and pay full price. For that you buy service, and with rejects often running at one third of any order, service is what you'll need.

Prepasted paper varies so in shrinkage, some paperhangers paint the wall behind where it is to go the background color of the paper, hoping the joints won't be too noticeable! Some wall conditions don't take to some papers, so you'll need expert advice. Designer papers can cost two thousand dollars for a bathroom, although they aren't necessarily printed or blocked any better than cheaper ones. Many papers' instruction sheets say, "Aim for the best [pattern] match at eye level," which means the pattern is apt to run differently from sheet to sheet.

Finally, remember that a good painter, or finish carpenter, or tile setter, or plumber likes to work on a good-quality house. You might well find you can afford the best all-around crew simply by specifying a simple but well-built house. In any case, your painter is a vital part of the whole, one who can make or break the result of all your efforts and money. Finishing is one more place to spend enough for a top-notch job, and enjoy the rewards for years to come.

17

Getting Done

WHILE THE SUBCONTRACTORS are finishing up inside, the builder's crew goes back out to complete the details of the exterior: installing gutters, outside lighting and steps and walks, landscaping, and any unusual pieces of trim or hardware. Much of what goes on from this point is subcontracted, inside and out, so the crew thins out and mostly just the foreman and a helper are at the site. It's important to have your builder or his foreman on the job right through to the end. He knows the work is done right, and you know he knows.

Gutters first. I just don't know which kind is best. Wood gutters, well cared for and securely fastened to a properly designed cornice, will last several decades. They cost the earth, and will therefore likely be carefully installed and maintained. A north side wood gutter rarely emptied of leaves will not only overflow but rot out long before a metal or plastic gutter in the same location. And I know of houses with eighty-year-old wood gutters that are still serviceable. For some house designs, there is no reasonable substitute.

I recommend lining wood gutters with roofers' membrane, and applying an oil-base paint or stain outside every two to five years. Wood gutters must be spaced about one-half inch away from the cornice, using pressure-treated spacer blocks. The air circulation here will allow the gutter to live out its full life. I use a 45-degree miter joint to splice wood gutters, apply silicone caulk, and screw each together. As with any gutter, keeping debris out is required maintenance.

The most common guttering is aluminum, called K-section, after the rough shape of the letter's upper half. This is available in lengths up to twenty feet from lumber yards, or in full house lengths from specialty subcontractors who roll-form the stuff right on the site. Gutter companies are often siding companies as well, and often operate right on the cutting edge of competitiveness, with rascals and incompetents at every turn. Especially on an older or poorly maintained house, installation is the crucial part of a gutter project, and a poor job won't show up until well after the check clears. Gutter intallers must decide on the spot whether that slightly rotten old cornice will hold the fasteners. If they're paid by the installed unit, they're unlikely to agonize much over the judgment call. Even on a new house, when speed is the byword, sturdiness isn't.

So what do you do? Have your builder either apply the gutters himself or directly supervise a sub's installation. Some builders like the vinyl or PVC guttering for its classic C shape and round leaders (vertical pipes). I like them, too, though the brackets are fussy and can't be used with fascia boards far from vertical. Aluminum gutter supports suspend the trough from the roofing, none too sturdy; hang from brackets nailed into the fascia, OK as long as the fascia is; or nail the gutter directly into the fascia and rafter ends with enormous aluminum spikes. The spikes often split the fascia board, and are worthless unless they hit a rafter tail. Of the three methods, the fascia brackets seem best.

The trouble with all this is that gutters in snow country take lots of abuse. Snow sliding down the roof rips at them. Snow and ice sit in them for weeks, straining the brackets with hundreds of pounds and pushing mightily at the joints. If you leave the gutters off, the foundation drainage must deal with lots and lots of water, as must the siding at the base of the walls. Damned if you do . . . More often than not, I recommend them, and favor mounting them below a line sliding snow might take. The cornice must be designed with this in mind, though, so it's just one more thing you must plan ahead of time.

Probably the main damage to gutters is done by ladders. Going up on the roof, or up to clean out the gutters themselves, puts pressure on the front edge of the trough. Sound wood gutters have no trouble with this, but aluminum and vinyl ones always give under the weight. Vinyl pops back into place, aluminum barely does. Either way, the ladder is not safe. Lean your ladder against the side of the building and use ladder arms, wide U-shaped brackets that hold the ladder out from the wall. Don't risk an accident and mangle your gutters by leaning on them.

Gutter leaders should dump their streams of water away from the foundation. If you included underground pipes around the house for gutter water, good for you. Fittings are available to connect the leader to the large pipe sticking up from the drain line. Make sure the drain pipe puts no pressure on the leader in any direction. If you have no drain lines, at least use splash blocks to get most of the water well away from the foundation. The ground around the house should slope away from the foundation, too, to educate the water flow. With properly sloped backfill it's also possible to bury a four-inch pipe about a foot below grade to carry gutter water away from the foundation. Make sure the leader isn't jammed into the elbow near the house, though, as this will move up and down with frost action.

Use one leader for every twenty feet of gutter. Fasten all gutter and leader joints with rivets or screws and sealant, except the top angle connectors, which you may have to remove to clean. Use leaf guards, small wire cages, at the top of the leader connectors. Pitch the gutters toward the leader connections at least an inch every fifteen feet or so. Make sure the hangers are rugged, especially in snow country. Don't fasten any bracket to a piece of wood of questionable strength, as it will surely fall off.

Next we might install shutters, if the design calls for them. Purists will want wood shutters, pragmatists vinyl, and nothing I say will make any difference. Wood looks nice and requires maintenance, difficult maintenance at that. Vinyl looks like vinyl and always will. The only caution either way is to avoid splitting the clapboards by drilling holes for the shutter brackets or screws. Incidentally, be careful when you grab your shutters to remove them at painting time; they make great wasp condos.

Any hardware or appurtenance should be fastened to the

house with due regard for a lasting job. We like to fasten a block of trim material behind almost everything that will be installed to the walls of the house. This includes electric and gas meters, hose bibbs (outdoor faucets), light fixtures, even flagpole sockets. The trim block makes a secure fastening point for the hardware, and ensures that the attachment will be flashed to turn water away. We form the top of the block with a lip to tuck behind the siding, or use a regular wood drip cap over it.

Landscaping and walks are often done well after the house is livable, or even lived in. That's because they are easy things to toss out in budget sifting sessions. One advantage of the postponement is that the backfill has time to settle before the work is done. The disadvantage is that you've done all the work and spent all the money and the place still looks incomplete from the street. That's the main reason spec builders landscape as soon as the siding's on; it dresses the place up. If you don't mind waiting, though, your reward will be a more lasting job.

Consider having no foundation plantings. These devilsome shrubs look skimpy when they're new and the price is ringing loudly in your ears. Then, when everything's well settled, they outgrow their allotted space and must be hacked to shreds. Otherwise they lay siege to the house, making painting and window washing hazardous, and mold and rot likely. Buy the bushes if you must, but plant them away from the house, and use a ground cover or even the dread grass against it. Though lawns seem to be out of favor, grass dries out quickly and doesn't harm the house. Whatever you do, *do not* use wood chips on the ground around the house. Ants love them, and quickly range around and into the house.

If you run the finish grade above the height of the basement windows, you'll need an areaway. This is a steel or concrete semi-circle that keeps the dirt away and fills up with debris. In a house with no gutters, the areaways may fill up with water and wreck the windows. This can be prevented by draining them using a pipe or coarse stone underneath down to the footing drains. Areaways can't be easily fastened to a foam-insulated foundation. I suggest using the concrete units and letting them just sit there unattached. They'll stay best if the backfill has settled some months before you place them.

18

Taking Care of Business

THERE ARE A FEW small jobs left, but your house is nearly done. Everyone who's worked on it is tired and ready to move on. Even you will have a hard time sticking with the project now, no matter that you're so close to moving in. Mercifully, weariness tends to smooth some of the rough edges of the final dealings. Don't put your feet up and break out the Jack Daniel's yet, though, for attention to the last details pays off. Your business with your architect and builder can't be completed until all the work is finished.

In order that all parties agree on just what "finished" means, you should write a punchlist, your version of what work remains. Most of what you list will be obvious to you and your builder, but set it down anyway. You might make out three or four of these informal documents before the job is completed. Date each one and keep your own copy. Be sure to include a final cleanup of the site and house, in the manner stipulated in the contract. You can ask your architect's advice about what to list, but write and present the list yourself. Finishing up is between you and your builder.

Talk over each punchlist item with the builder or foreman. He will most likely understand what needs doing, unless the list contains a change from the contracted work. Changes mean delays, always, and also may mean you'll have to involve the architect. If your punchlist asks for work beyond what the contract specified, you may not get it; builders' schedules commonly are tight and haywire at once. If your builder refuses to do work that *was* stipulated, you'll have to use whatever combination of carrot and stick you think will move him. In any case, discussing the tasks in some detail will confirm that each person is aware of the other's expectations.

Rare is the job that doesn't contain some miscommunication or bruised feelings; this is too complex and emotional an undertaking for that. After all, yours and your builder's financial motives run in different directions from the time the contract is signed. It's great if you end up friends with your builder or crew. You'll help make that possible by concentrating on keeping your business relations open and fair.

Don't start handing out punchlists until the crew seems to be looking around for things to do. No one likes to be reminded of the task he was about to start anyway. I usually figure punchlist work takes two or three weeks. You needn't anticipate how easy or hard it will be to complete the items on the list. Some will undoubtedly involve rescheduling subcontractors, especially the painter, but that's the builder's concern, not yours. On your part, try to be understanding if the punchlist work takes some time.

You might, thinking to give yourself leeway, tell your builder you must move in on August 1, though you can actually wait until September. This tactic might hurry him along, but only until August. Then, when you're clearly *not* moving in, and the crew realizes it was being manipulated, you'll get little cooperation. Straight dealings are best for all concerned.

By the same token, avoid moving in before all work is completed; doing so makes any remaining tasks more difficult. Possibly some of your belongings will be damaged. You'll have less time to cope with the stresses and logistics of moving. Worst, perhaps, your being in the way muddies your final business dealings with the builder, and excuses some poor behavior from both parties. Rent a house, move in with friends or relatives, or go on an unscheduled vacation, but stay out.

When everything is installed and fit and painted and cleaned up, the job is done. There'll probably be no definite point at which this happens, no champagne bottle breaking across the bow. You and your architect and builder will agree that everything's substantially finished, and that's that. It might take some time to coax the final few licks of work from your builder, since he's on to his next house. Keep after him, though, for it's in everyone's interest. He may be reluctant to hand over this gleaming product of a chunk of his life to its rightful owner, but he can be convinced.

When the punchlist and the moving in are behind you all, have a party. Human beings like ceremony, and settling their differences with food and drink. Invite everyone who worked on the project, if you are willing. There aren't many rules for such a gathering, though matters of business should be left to another occasion. It's time to step back and admire your addition to the world's landscape, and celebrate the start of your new life at home.

Index

Page numbers in italics refer to illustrations.